A GUIDE TO C
WEIGHTS AND

articles, tables and notes, to define, explain and comment upon the modern uses and relationships – as well as the global origins, evolution and culture – of the units composing what became known as the Imperial System, but has since become the universal Anglo-American system of customary measures

for use in the home, in schools, by government and public bodies, by trade, industry, the media and in countries overseas

for quick reference, occasional consultation, continual study or simply reading for pleasure

compiled and edited by Vivian Linacre *on behalf of the*

BRITISH WEIGHTS AND MEASURES ASSOCIATION

45 Montgomery Street, Edinburgh, EH7 5JX   tel/fax: 0131 556 6080

web sites: www.footrule.org   www.bwmaOnline.com   *http://users.aol.com/footrule/*

2001

Dedications

to

Leo Blair

because

he, and the first draft of this book, both arrived on 20 May 2000

his initials spell the symbol (lb) for the pound weight

the immediate announcements from Downing Street, every newspaper report in the days following, and the congratulatory resolution of the House of Commons moved by the Deputy Prime Minister, all referred to the baby's weight at birth as '*6 lb 12 oz*', despite the fact that his father was head of the government which – as from the beginning of that year – had made the primary use of Imperial units of weight and measure a criminal offence

and – more seriously – to

all other young people who – in the name of 'political correctness' or 'progressive education' – are deprived of this cultural heritage and wealth of useful knowledge

'Stand up for the foot'

*(Chris Jagger: from the album **'Atcha'** – 1993)*

I like to measure my speed in miles per hour, when I'm travelling down the road
If I go to a bar I drink my beer in pints, I use tons for a heavy load

So don't give me kilometre, centimetre, millimetre, parking metre, none of that crap
I am a man who stands six feet tall and wears a ten gallon hat

Stand up for the foot, it's feeling the pinch, stand up for the foot and don't you give an inch

I don't mind a litre of French wine but don't tell me how to think,
If I want to buy me a pound of potatoes or put a gallon of gas in my tank
My feet are size nine, I can dance down the line to gain me another thirty-two yards,
But tell me I've gone thirty metres and I'm bound to kick your ass

Stand up for the foot, it's feeling the pinch, standup for the foot and don't you give an inch

You won't be reaching for a big ten-incher, be a member of the 'mile high club',
Or go out looking for a pound of flesh and having an ounce of luck

So don't give me kilometre, centimetre, millimetre, parking metre, none of that crap
I am a man who stands six feet tall and wears a ten gallon hat

'A Guide to Customary Weights and Measures'

Copyright © VIVIAN LINACRE 2001   ISBN 1-872970-34-6

Published in association with Winter Productions Ltd

Price (incl. postage) £7.00, or £18.00 for 3 copies, £30.00 for 6, £48.00 for 12

Cover design by Stewart Buchan

Printed by Paramount Printers, Edinburgh.

PREFACE

The British Weights and Measures Association has for several years received enquiries, at an ever-increasing rate and from throughout the world, for a 'Guide to Customary Weights and Measures' in a popular, up-to-date and convenient form – enquiries which could never be satisfied because apparently no such publication is available. It is hoped that this volume will fill that gap.

Of course, collections of tables are produced, but without the necessary explanation of relationships among the sets of units for the various measures within the system, and without the historical and cultural background that gives the system so many extra dimensions. There are also several excellent compendia and works of a technical nature, analyzing imperial units within specialist mathematical, commercial or archaeological contexts, as well as comparative studies detailing imperial measures in relation to the metric and other systems from different countries and ancient civilizations. Yet none of these meets the essential requirement; for an introduction and companion to the old system – comprehensive and authoritative yet simple and practical, educational yet entertaining – designed for a mass readership.

Many people in Britain, who were educated in our system and remain indifferent or even averse to the use of metric, complain of the lack of a handbook on imperial from which to refresh and enlarge upon their recollections and to pass the knowledge on to others. Many others, who are more or less familiar with both imperial and metric, wish for a proper understanding of the former to relieve the current confusion. Everyone else, educated only in metric, needs to know far more of the vernacular system that they are using for everyday purposes – and which the world's most popular children's authors of our times, Roald Dahl and J K Rowling, use exclusively in all their books! Above all, there is a desire to learn more of their traditions and cultural identity.

There is a huge potential demand from overseas, too, especially in the USA – the world's super-power – where the 'inch-pound-quart' or 'English' system still predominates, as well as in our former Dominions, most of which have gone through the trauma of metrication though some remain ambivalent. Finally, this might even serve a missionary purpose in those countries that have known nothing but 'metrickery' for the last century or two, to let them see what they are missing!

Naturally, the beginning of measurement was *time* – time for measurement of the seasons and for study of the heavens that ruled early mankind's lives – and time for measurement of music, since music is based on time and was pre-literate society's medium for transmission of knowledge as well as for ceremonial. Next to be learnt was the measurement of *length* – for timber to build shelters and great stones for temples, and the measurement of territorial boundaries and distance for journeys. Then people, or the priestly caste, mastered the calculation of *area*, to measure fields for cattle and crops, with the establishment of agriculture. They next learnt to measure height, to measure *cubic* capacity, from which followed the measurement of fluids. So, finally, evolved the skill to measure the *weight* of volumes of wine and water and the weight of produce for trade. That was the natural sequence – the necessary order governing the gradual application of the *nature of number* to man's working environment. That relationship and process are the soil and seed of our customary system of weights and measures.

The study of metrology, especially its history, attracts mystical speculation which, to the extent that it stimulates the imagination and opens the mind to fresh connections is all to the good but, insofar as it tends to select or massage facts to suit fantastical theories is wholly bad. This publication prefers to sustain a sense of fascination and wonder, rather than take the cynical, repressive option. So some controversial material is included to excite the reader and illustrate the infinite range and ramifications of the subject, despite (express or implied) reservations as to its scientific validity. Generosity of spirit can, after all, be combined with a healthy scepticism. Compared to New Age cults and the vogue for alternative systems of belief, anything here must seem fairly orthodox. Form your own judgement, but please notify us of any apparent errors or oversights – for which I alone am responsible – so that they may be corrected in a second edition.

I must record my debt to the several named contributors, and gratefully acknowledge invaluable help from John Strange, advice from many other colleagues, especially BWMA's Chairman, Bruce Robertson, Robert Carnaghan, Bill Peters, Antony Flew, Arthur Whillock, Simon Hooton, Stephen Fenn and John McLean, encouragement from several of our Patrons and Honorary Members, as well as from Seaver Leslie of *Americans for Customary Measures*, constant inspiration from John Michell and John Neal, and blessings from my collaborators on the late Anne Macaulay's book.

Vivian Linacre (Director)

Principal Contents

Introductory Articles	1
STANDARD LINEAR MEASURES	10
NAUTICAL MEASURES	15
Light second, Light foot, Astronomical unit, Light-year	16
Bed sizes, English shoe sizes, Biblical lengths, Numbers used in trade	18
Deriving from the human body, Equivalents of 1 inch, 1 foot, 1 mile	19
Surveyors' measures, Decimal conversions	20
CIRCULAR MEASURES, WEIGHT AND MASS	21
Avoirdupois WEIGHTS, TROY WEIGHTS	22
APOTHECARIES' WEIGHTS, Maundy pennies	23
MEASURES OF AREA IN EVERYDAY USE	23
Gunter's square land measure, Builders' square measures	24
MEASURES OF VOLUME AND CUBIC CAPACITY	24
FLUID MEASURES, UK and US fluid measures	25
In the Kitchen, Wine and champagne, Beer and ale	26
Apothecaries' fluid measures, Medicine measures	27
Paper sizes	28
Paper folding into leaves	29
Sizes of bound books	30
MUSIC	31
SOUND	32
PRESSURE, TEMPERATURE	33
FORCE	34
WORK and ENERGY	35
POWER	36
SPEED	37
AN ANTHOLOGY	37
Megalithic measures	42
Cosmic numerology and customary measure	44
EPILOGUE (the metric alternative?)	49
Appendix I (foreign units corresponding to the foot and the mile)	55
II (quotations from Shakespeare)	56
III (Committee Reports 1816-1821)	59
IV (John Quincy Adams)	62
V ('The Gallon')	64
VI ('Rules of Lawn Tennis' and 'Proof Spirit')	66
VII (The 'King's Girth')	67
SELECT BIBLIOGRAPHY	75

Introductory Articles

From ***William D Johnstone's 'For Good Measure' (NTC Publishing, USA, 1998)*:** "Any number is generally one of four types: *Decimal* (divisible into tenths, a system handed down by the Chinese and Egyptians); *Duodecimal* (divisible into twelfths, a system tracing from Western European pre-history and commonly used in the Roman Empire); *Binary* (divisible into halves, quarters, eighths, etc, originally Hindu, but fundamental now to computer mathematics); *Sexagesimal,* divisible into sixtieths, originally Babylonian and reflected today in the measurement of time and in geometry. [see *footnote A*]

With all systems of measure, certain basic definitions apply. *Fundamental units* are quantities whose scale of measurement is arbitrarily assigned and is independent of the scales of other quantities; all other quantities are *derived units*, subordinate to the original fundamental units. A *unit of measurement* is a precisely defined quantity, in terms of which the magnitudes of all other quantities of the same kind can be stated; whereas a *standard of measurement* is an object which, under specified conditions, defines, represents or records the magnitude of a unit."

The transatlantic dimension must be introduced at the outset, for the imperial system has become the Anglo-American system. Indeed, the main purpose of compulsory metrication is to deprive Britain of what the EU sees as our "unfair competitive advantage" by sharing a common system of weights and measures with the world's superpower, to which UK governments respond: 'Quite right – it is iniquitous that Britain enjoys the colossal commercial and cultural benefit deriving from this common system of customary measures, which clearly must be abolished!' There are a few, interesting differences between US units and ours, but the origins and system are the same; just as we share a common language despite minor differences in usage and spelling. Indeed, just as more people in the world speak American English than British English, so there are far more Americans using 'English' units of weight and measure than British using Imperial units.

Thus, in the British system, the dry capacity units and liquid units are the same, whereas in the American system they differ. In Britain the bushel was always used as a unit of capacity for both liquid and dry quantities, but in America is a dry measure only. [see FLUID MEASURES] Also, the UK liquid pint and dry pint each equal 568ml; about 20% larger than the US liquid pint of 473ml and slightly larger than the US dry pint of 550.61ml. (Johnstone actually refers to '0.568 cubic decimeter', etc, but that would mean little even to fanatical UK metricists!) To be precise, the UK pint = 568.26125ml and the US pint = 473.17647ml. Again, the common US *short ton* = 2,000 pounds and the British *long ton* = 2,240 pounds. Furthermore, in 1893 Congress decided that the metre should be 39.37 in., making the inch very nearly 2.54000508cm. In 1897, the metre was legalized for trade in Britain; its value set at 39.370113 in., making the inch very nearly 2.53999779 cm. In 1959 the international inch was defined as 2.54 cm. So 499,999 US in. of 1893 equalled 500,000 international inches; a difference of 1 inch in nearly 8 miles, which is negligible. But the US Coast and Geodetic Survey still use the 1893 inch. All is explained later!

However, the assumption that weights and measures must be standardized by fixed units is comparatively recent, deriving only from the concentration of power in the person of the sovereign at the expense of both the landed nobility and the liveried trade guilds – each of which throughout the Middle Ages defined and regulated their own customary measures to benefit their respective interests. But the creation of a central government, emergence of a rapidly growing middle class, overseas possessions and a Royal Navy to protect them, all demanded a national code of weights and measures. Furthermore, the Crown, struggling both constitutionally and financially to retain control over this modern state, had few sources of revenue apart from excise and other duties levied on goods – both domestically produced and imported – for which, however, there was a vastly increasing potential as manufacturing and overseas trade expanded rapidly. So fixed standards of weights and measures became imperative, in order to collect and account for the appropriate dues on a uniform basis across the country.

It is fascinating that previously the chief economic concern had been to ensure the integrity of the currency – always threatened by debasement and counterfeit – whereas standards of weight and measure were of minor importance and accordingly allowed to vary geographically and fluctuate periodically; but gradually priorities were reversed as stability of the means of exchange had become assured whilst stability of weights and measures was sorely needed. That preoccupation with the specification and regulation of a national system of weights and measures has intensified ever since, whilst concern for the credibility of money has declined! We should realize that this was not always so – that standardization of weights and measures is not necessarily sacrosanct and in earlier centuries was

very properly regarded as of far less importance than safeguarding the value of money. Indeed, but for the fiscal plight of declining monarchies in Britain and Europe, and the profligacy of succeeding bureaucracies, protection of the pound sterling would have been given rather more attention than playing politics with the pound avoirdupois.

footnotes

A The last three of the four types of number – *Duodecimal, Binary* and *Sexagesimal* – are all compatible with the imperial system and with one another. Incidentally, one of the few early civilizations that used decimal counting was the Phoenician, whose merchants employed base 20, making a notch or *score* on a wooden stick when that number was reached. The Bible frequently refers to a score, and the French call 80 *quatre-vingts*.

B The House of Commons has recently produced its own style of ruler, suitable for slipping into MPs' brief-cases. A handsome design, emblazoned on one side with Parliament's logo (appropriately the 'Traitor's Gate'), it measures 30cm. in length, with each edge marked identically in cm.- mm. divisions and the back completely blank. That is all that a metric ruler can offer, just one unit divided by ten and ten again (in fact, of*ten*!). The official responsible was asked two questions. First, how can this length of 30cm. – itself not an authorized metric unit – be divided into a metre, which is the foundation stone of the entire metric structure? Answer: "unfortunately, it can't" (i.e. 3.33 recurring). Second, why 30cm? Answer: "It was as close as we could get to a foot" – shooting himself in the foot! 30cm. = 11.811 inches. So our rulers cannot even produce a proper ruler! [see EPILOGUE (4)]

C Our customary measures are indeed very largely duodecimal, because 12 is far more easily factorized than 10; being divisible by 2, 3, 4, & 6 instead of only 2 & 5. It also lends itself to halving and doubling [see eighth para. of the John Neal extracts below, and end para. under MUSIC]. As Stephen Fenn points out, financial markets throughout the world quote and calculate interest rates in fractions; from halves to quarters, 1/8ths, 1/16ths to 1/32ths. Decimals are inhuman, for the mind naturally thinks in fractions. If a child in Paris or Berlin is asked its age, does it answer "I'm 8½" or "I'm 8.5"? Are the men wearing no. 11 and no.14 rugby football jerseys 'wing-three-quarters' or 'wing 0.75s'? Do you agree to meet a friend in a ¼ hour or in 0.25 of an hour? As discussed later, not only are the clock and the calendar duodecimal, but so are the measurement of the Earth and celestial rotations.

from John Neal's monumental work, 'Opus 2 - All Done with Mirrors':

"It is pointless to start at the chronological beginning of a study of metrology: to all intents and purposes there isn't one. Mathematical skills of some degree are evident among societies that were believed to have possessed only the most basic necessities for survival. At Ishango in Zaire during the 1950s, a bone was discovered, which was shown to be between 9,000 and 11,000 years old. It is squared off and perfectly straight, about nine inches long and resembles a handle with a sharp piece of quartz embedded in the hollow end that was perhaps used for scribing. Scored upon its sides are a series of lines arranged in distinct groups. The archaeologist finder of the bone, Jean de Heinzelin, interprets the markings thus: in the first group the lines are 11, 13, 17 and 19 in number which he states are the prime numbers in ascending order. In the second, 3, 4, 6, 8, 10 and 5 then 5 and 7, which he believes suggests duplication by two coupled with decimal notation. The third group are 11, 21, 19 and 9 which he sees as 10 and 20 plus one and 10 and 20 minus one. He believes them to be tables of some description. Others maintain that it is simply some sort of tally or lunar count. Whichever is correct, the presence of the bone implies pre-historic numeracy.

Neither can one begin by trying to explain ancient metrology from the point of view of an evolutionary development; because the earliest known measures prove to be part of a completely developed system that on analysis is extremely sophisticated and identical in every respect to numerical systems used, the evidence suggests, universally. It would seem to be linked to the fully developed cultures that arose in the ancient world, also with no discernible evolutionary stages……

The modern search for a definition of the ancient measures begins in earnest with Isaac Newton. Before Newton, a Europe founded on Rome and [until the Reformation] dominated by the Church: after Newton, a world founded on science and dominated by industry and profit. Before Newton, there was little necessity to try to define ancient units of measurement as there was no application for them, nor would the implications of their geodetic relationships [*i.e.* relating to measurement of the surface of the Earth] have been understood, as no reliable measurement of the world existed. Newton believed, by tradition, that the 'sacred cubit' of the Jews was a geodetic measure and, although he anxiously sought an accurate estimate of the Earth's dimensions to substantiate his laws of gravity, he also attempted to deduce this from the analysis of ancient measurements. The modern objective search for these definitions can be said to have begun with Newton and continued to the present day.

That a circle has been measured since time immemorial as 1,296,000 arc seconds is truly remarkable, inasmuch as it is an extremely fine measure. (1,296,000 = 60 x 60 x 60 x 6) It is not unreasonable to conclude that the only reason for possessing a measure this fine would be for surveying on a geographic scale, and the fact that one second of arc of the meridian degree is 100 'Greek feet' is no coincidence......[But while the circle has been divided into 360° since pre-history, there is scant evidence that the degree was divided into 60 minutes before the last millennium BC, nor that the minutes was divided into 60 seconds much before the Renaissance. Moreover, the ancient Greeks could not have known that the earth was not a perfect sphere (Hence the slight variations in 'Greek feet'?). One second of arc, measured along a meridian, is about 100.77 British ft. at the equator, 101.02 ft. at a latitude of 30° and 101.53 ft. at a latitude of 60°.]

Newton did not know that in 1664 one Richard Norwood, author of *A Seaman's Practice*, had taken observations of the sun at York (probably with a backstaff, since the sextant was invented by John Hadley in 1730) and another at Tower Hill in London, computing the degree to be 69.5 statute miles. Whether this would have been a sufficiently accurate computation for Newton's purposes must remain academic, he being unaware of it. His requirement of an accurate assessment of the length was finally satisfied by the efforts of a French astronomer, Jean-Luc Picard in 1671. His measurement of 69.1 English miles to the degree was not bettered for the next 150 years – a computation good enough for Newton to finalize his general theory of gravitation. The correct figure would have been between 69.114 and 69.115 miles. The average per degree is 69.055 miles, very close to the measure of 69.06 miles between 45°

and 46°. [So the British nautical mile is accurate at about 48° N – see NAUTICAL MEASURES]

Probably the first reference to measures as statutes are those of the laws of Athelstan, who became king of Mercia on the death of his father Edward the Elder, in 934. The following year he became the first elected king of all England and was crowned at Kingston. His first tasks as sovereign were to establish the boundaries of his domain and institute just laws. [see APPENDIX VII] His law reveals that the measures used by Athelstan, probably inherited by him from remote antiquity, and later defined by the proclamation of Edward I in 1305, were identical to those we use in the present day.

Philologists claim to have largely reconstructed the ancient vocabulary that indicates a vast commonality among ancient peoples. Countless nations may as well have poured upward from a hole in the earth, for all that is truly known of their origins. What has become apparent is that although their language may have mutated, their relatively uniform measurement system did not. It has not only been shown to have remained in continuous use from the prehistoric ages to the present day, throughout the lands populated by Eurasian peoples, but beyond any boundaries that we believe they could possibly have crossed. Yet virtually nothing is known about this concrete evidence, while – in contrast – scholars interminably discuss mere hypotheses concerning the origins of words and racial migrations!   The dissemination of the universal metrological system has been merely touched upon here. Just enough evidence has been presented to establish the general theory. So much is yet to be investigated: nothing has been mentioned of Angkor Wat, or even the Sumerian ziggurats, while the great pyramids and temples of China have received no mention. Tiahuanco in Bolivia is ripe for measurement, and a thousand other ancient monuments and forgotten cities from the Andes to the Gobi await the theodolite and yardstick to peer into their designs.

The ancient system has nothing to do with quantifying, although it may be used as such. This may seem a contradiction in terms – non-quantitative mathematics – but an enormous amount of mathematics is actually non-quantitative. It is far more of a set of proportions than a means of calculating amounts. The system goes far beyond being based upon nature in that it is simply geodetically harmonious. The very means by which it divides by halving and increases by doubling is imitative of such natural processes as cell division. Its counting bases more resemble those of the modern computer binary language: 16 bits, arranged in nibbles and bytes,

a kilobyte being 1,024 bytes and 16,384 bits – all on base 4. Counting by tens expresses only the quantitative aspect of both ancient and modern systems, that are then applied to an underlying harmony of pure number, which has other means than the decimal by which to propagate.

How the binary system operates, may equally be regarded as a unitary system, for it is simply based on the number one and the absence of one, or zero. How very Pythagorean – 'All numbers are in one, and one is neither odd nor even'! Then, regarding metrology, this simple number is given a linear value, which is the English foot. Everything else may then be given its proportionate relationship and a means of comparison, and this has been seen to extend far beyond measures. It has spilled into ratios.

Yes, there *are* 252,000 *stadia* or 18,000 miles in the circumference of the Earth, there *are* 210,000 cubits in a meridian degree, man *is* the measure of the world, and the cosmic model *can* be constructed from a Pythagorean triangle. The same light that shines forth from the crowded monuments of Egypt illumines stone circles on lonely hills in Scotland. Throughout Asia and the Americas, that same light is wherever you care to look – ruined, yet undiminished. Although the basic premise, prehistoric science, may be anathema to the prevailing orthodoxy, at least there are no institutions today that would actually brand the content as blasphemous or seditious. Yet with the dawning of the year 2000, these ancient standards were declared illegal to use as trading standards in Britain, the country that had nurtured and propagated them!

Probably for the first time, this knowledge is revealed into a world that has forgotten it even exists. For it goes far beyond metrology, into the number system that governs nature or physics. It was not *devised*. Too many points of correspondence indicate that the system was a *discovery*. It is no more an invention than Newton *invented* gravity or Einstein *invented* relativity – all they did was to devise mathematical models to explain natural phenomena – but this knowledge was understood in the ancient world with as great a clarity as those subjects are understood in the modern."

from Stephen Strauss's 'The Sizesaurus'(Key Porter Books Ltd, Canada, 1995):

"Merchants in mediaeval Danzig [now Gdansk] would cloak inflation simply by changing measures. Thus the size of a purported 'pound's worth' of bread changed, depending on the cheapness of the wheat going into it, but the price stayed the same. One should perhaps pause and reflect about this other way of doing things for a moment…..if some mediaevalist of a merchant tried to institute the elastic pound into his dealings today, his customers would think him not quaint but deeply immoral. Price can yo-yo up and down, and inflation can explode into hyperinflation, but in our minds pounds…[etc.] are immutable. We imagine that to change them is in some way to change the laws of nature, and yet, only two hundred years ago, it was not clear at all that the only way to construct an economy was to keep the weights and measures fixed and fiddle with the money. Not clear at all.

But the most striking feature of metric's conversionary process has been the resistance of the old religions. Even though England had been asked to join at the very beginning, and the American *philosophe* Thomas Jefferson proposed a decimalization of the English weight system in 1790 [later, in 1806, he rejected the metric system because it could not provide for conversion between time and distance measurement], an ecclesiastical war followed the birth of metric.

Initially, much of this undoubtedly reflected a continuing revulsion against the excesses of the French Revolution and its deification of reason over religion. The ten-day-week metric calendar had not made provision for a Sabbath, and Christians in the United States in the nineteenth century found this proof that the system was essentially an abomination. [see **Napoleon** in *An Anthology*] We are all one weighing-and-measuring church, but that church's pews do not provide an altogether comfortable fit. Metric is uniform, and its rigidity has foisted a uniform imperfection on the world. A gram is really too small to weigh anything but a single paper clip or a large raisin. Counting by tens is of course a rational way of measuring but…..it often flaunts its precision in face of the lower level approximations of everyday life.

Asking for a pound of sugar seems more natural than requesting 500 grams of anything. Doubling or quartering of measures appeals to something quite primal in the human brain.* It ties into mental arithmetic…..Metric eschews vulgar fractions and, in so doing, hides another truth…..And that can be demonstrated by what everyone recognizes as the most common mistake in applying a metric measure. The easiest stumble is to be off by an order of magnitude (by misplacing the decimal point). Can anything be more odd? In the name of a universal precision we have allowed ourselves to be ten times more dramatically wrong than in the past. The vision of an absurdly inaccurate accuracy seems a fitting end to this little reflection on the two-hundred-year revolution in measurement and the ultimate installation of the metric god."

[*Being born with two hands and two eyes, we innately conceive of a weight being held in one hand or the other, then half in each hand; which is why supposedly metric weights are forever halved and quartered, irrespective of the decimal units – see *footnote C* above]

STANDARD LINEAR MEASURES

1 foot = 12 inches   1 yard = 3 feet   1 rod, pole or perch = 5½ yards
1 furlong = 40 rods = 220 yards   1 mile = 8 furlongs = 320 rods = 1,760 yards = 5,280 ft

From *Measure for Man*, an article by the engineer and historical metrologist, Arthur Whillock, published in *The Dozenal Journal* (1994):

"An appropriate unit of length is, of course, the basic requirement for any measuring system, so an anthropo-compatible range was an inevitable beginning. Names for components of this were transferred generically to comparable sizes as more definite means of determining these were found and simple relationships defined: some became more important than others, notably the Foot. There were two forms of this. (a) The natural or ploughman's foot of twelve digits – about 9½" – which in early days was half the Egyptian natural cubit or forearm-length of twenty-four digits; this was used by Celts and survives today in Wales and Scotland. (b) The visual foot, towards which derivations by many methods have tended, is the more practicable by being the widest spacing that is comfortably perceived without head movement and therefore the most efficient to use for a hand-held measuring tool.

Visual feet [see APPENDIX I] range from 13.2" for the Indo-Aryan or Northern foot, described by the great Egyptologist Sir William Flinders Petrie (1853-1942) from a study of early buildings before an actual specimen was found, to 11.53" for the value of a Roman foot as used in southern England. The Romans took their foot from an early Greek unit of sixteen digits, 11.8", which they divided into twelve *pollices* (thumbs) for incorporating into their *uncia* system of weights and measures. Larger, 14", feet, found in such diverse areas as Poland, Italy and China, were probably half a *Braccia* or arm-length and favoured by cloth workers, an obvious measure for their purposes. The English measures common in the cloth trade until recently were the *nail* of 2¼" – i.e. $1/16^{th}$ of a yard – and the *ell* of 45" or 20 *nails*.
William of Malmesbury stated that our yard was a distance given by the outstretched arm of Henry I (1106-1135) from nose to finger-tips, which must have been the "Iron Ulna of our Lord the King" as mentioned in mediaeval

documents. True or not, it was an appropriate way of fixing a length that all could appreciate – now affording an Aunt Sally for intolerant metricists who can only mock at simple things that they cannot understand.

There is no definite evidence on the origin of the English foot [but see John Michell's references below: also in *Megalithic Measures* and APPENDICES I & VII], traces of which can be found in old buildings back to the 10th century. Both the Roman foot of 11.53" and the Greek common foot of 12.45" were used then for building work, so it has been suggested that our foot, which is known and used world-wide, was an average of the two, but units of length did not evolve that way. Many documents as far back as Saxon times give the length of a foot as *duodecim uncias pollices* (twelve thumb inches) It was thus a true anthropometric measure that Edward I (1272-1307) legalized at three barleycorns to its inch. Flinders Petrie deduced that our foot has remained unchanged within 0.5% since then, being based on principles acceptable to the users. (I wonder how long the metre would survive without the aid of elaborate equipment and legal sanctions?)

Of special interest, too, is the Greek *Olympic* foot, used for the layout of the Parthenon. This is a geodesic unit that divided a minute of arc on an assumed spherical Earth into 6,000 parts, with 600 to a *Stade* for one tenth of a minute. This is a controversial theory, but the parallel with our nautical mile is remarkable, as with the original measurement for the metre, which is 0.2mm shorter than was intended. The Egyptians estimated the minute as 5,000 of their *Remen* units, so some authorities assume that the Greeks obtained their Olympic foot from them, but the sexagesimal format indicates a Sumerian origin. Accurate measurements like these would be possible on extensive river flood plains, with angles obtained from sun shadows or star sightings.

Human-sized measures are essential to allow correct perspectives toward our daily affairs. Just as a valid unit of length complies with visual acuity, so weight is judged by our muscular reactions to it. [Weight is a force, mass is not.] Piles of cobbles found at Neolithic sites were evidently the camp arsenal, ready to discourage prowling animals or repel unwanted visitors. These were no doubt selected for optimum range, accuracy and effect, with median weight somewhat less than one pound, appropriate for use by short stature people. (The 'war stones' collected by Pitt-Rivers from Polynesian natives were heavier.) It is not too extravagant to suggest that an acceptable modern unit of weight should be a direct descendant of early throwing weapons. Cereal seeds, the units of settled living, by being readily available and appearing regularly in nearly the same sizes and weights, were widely used as reference standards for both."

from: '*The Ascent of Man*' by Jacob Bronowski (BBC 1973): "The largest single step in the ascent of man is the change from nomad to village agriculture. What made that possible? An act of will by men, surely, but with that, a strange and secret act of nature. In the burst of new vegetation at the end of the Ice Age c.10,000BC, a hybrid wheat appeared in the Middle East…..wheat and water, they make civilization."

Edward I specified (*Statutum de Admensuratione*) 3 barleycorns for his English inch, 12 of which became the European standard foot; in contrast to the old Northern foot, which comprised 10 inches, measuring 5 wheat grains end to end – giving rise to the longer Saxon foot and its yard, and subsequently to the legal value for the Rod, Pole or Perch measuring 5 Northern yards or 5½ standard yards*: hence the quarter chain, 40th furlong or 320th mile. [*This 11/10 ratio recurs below.] As Whillock explained, Edward's new measure was essentially a 'Craftsman's foot', for use in building –"everything from cottages to cathedrals" – whereas the old Northern foot, yard and rod were primarily for land measurement.

Ian B Patten of Anchorage, Alaska, published an article in *The Kingdom Digest*, Dallas, Texas, reproduced in *The Voice of Avalon* (No. 26, Winter 1995), that asserted: "To have any value, a land measurement must reflect a compatible time-distance relationship. The Ancient Egyptians reckoned the Earth's rotational velocity at the Equator is 1,521.75 feet per second, and made their cubit one thousandth part of this: 1.52175 feet or 18.261 inches. So every four seconds the Earth moved 4,000 cubits which is one nautical mile, equalling one minute of arc. Thus, there was total harmony between distance and time through the cubit as there is with the nautical mile today."

This shows how far-fetched are so many theories in this sphere – especially the Egyptological aspects. In fact, the earth's equatorial radius is 20,925,646 feet and the earth's angular velocity is 72.92115 microradians per second, so the rotational speed at the equator is 20,925,646 x 72.92115 x (10 to the power minus 6) = 1,525.922 feet per second.

He goes on: "Even the Anglo-Saxons were equally as astute in metrology as the Ancients. They had a dual system: linear units for straight lines – the digit (0.72"), the fathom (72"), the inch and the foot – and those for land measurement, which bore a 11:10 ratio to the straight line units.... Thus the link was 7.92" which equals 110/10ths of the digit; while 10 links (79.2") equal 11/10ths of the fathom. The 100 links of the chain of 22 yards and the 1,000 links of the furlong (1/8th of a mile) make the sides of a rectangular acre of 43,560 square feet. And the 640 acres in a square mile was always so easily divisible into sections of 320, 160, 80, 40, 20, 10 and 5 acres....Furthermore, the Anglo-Saxon rod of 79.2" is exactly **double** the *true* measure of the metre at 39.6", therefore equalling one ten-millionth part of one *half* of the Earth's circumference and hence coinciding with the 12-hour clock. Again, this *true* metre of 39.6" is in exactly the 11/10 ratio to the Imperial yard…Hence, the metre fixed at 39.37" is an abomination, almost a quarter inch short – over nineteen feet short per kilometre. (Even the earlier French system was more scientific than the metric; for the counterpart of the fathom was the *toise de Perou* standard, which had 864 *lignes* in its six *pieds*. 864, of course, is one hundredth of 86,400 which is the number of seconds in 24 hours. So when metricists dismiss our duodecimal system as an antiquated hotch-potch they should look at the facts, which are that the Ancients and Anglo-Saxons knew far more about mathematics and metrology than a mob of revolutionary hot-heads a mere 200 years ago!)"

Many of Mr Patten's sidelights are illuminating, but he illustrates too clearly that numerology is to mathematics as astrology is to astronomy! In particular, there is no real evidence that the earth's polar flattening was even suspected prior to Newton. Besides, as John Michell has pointed out, there is no real relationship between the 79.2" rod and the Earth's circumference, which measures in ***miles*** exactly one tenth of 12 to the power 5 (see *Cosmic Numerology*). Likewise, the circumference of the Moon measures in feet exactly 12 to the power 7. This, indeed, is the principal standard of astronomical measurements. **It is vital to understand that geodetic values are not derived from actual measurement but from an integral set of fractions: all of them fitting together by simple ratios. As Michell remarks: "The whole thing is a pre-existing number code, beautiful and subtle, which is surely the prototype of Creation."**

[To come down to earth with a bang once again, it must be pointed out that the earth's (maximum) equatorial circumference is not quite 131,479,700 ft, half of which is only 9,960,584 Anglo-Saxon rods of 79.2". So this rod is too big by about 4 parts per thousand for even the earth's greatest circumference, whereas the metre is too small by only 2 parts per 10,000.]

In 1861 Sir John Herschel, following his calculations relating to the Earth's polar diameter (axis), suggested a new "earth-commensurable" unit of linear measure which he called the *geometrical inch*; exactly one thousandth of an inch longer than the Imperial inch (1.001") and representing one 500 millionth part of the Earth's polar diameter. [But the polar axis is about 500,531,700 inches, so 1.001 is not quite enough: 1.0010634 would be closer] Herschel explained, however, that by this tiny adjustment, a geometrical half-pint is exactly one hundredth part of a geometrical cubic foot and a geometrical ounce is exactly the weight of one thousandth part of a geometrical cubic foot of distilled water. [But 1 cubic foot contains 49.8306836723 pints, and if that number is multiplied by 1.0010634 *cubed*, we still get only 49.9898, not 50.00!] Curiously, according to many Egyptologists, this same unit of 1.001" is the key to deciphering the Great Pyramid. Furthermore, considerable mediaeval evidence suggests that the former standard linear measures were very slightly larger than the modern, the original standards having been lost in the House of Commons fire of 1834 – *not,* as frequently stated, the Great Fire of 1666!

Eratosthenes (c.276-193BC) is generally credited with being the first to measure the Earth's circumference. When the Sun was vertically above Syene (modern Aswan), at mid-day at the summer solstice, it was 7°12 min. from the vertical at Alexandria, where Eratosthenes was librarian. The two places are about 500 miles (5,000 *Stadia*) apart, but not quite on the same meridian; so he calculated the Earth's circumference to be 500 x 360 / 7.20 = 25,000 miles. This compares with the true distance of 24,900 miles – polar circumference 24,860 and equatorial 24,901 miles – not bad! [see below] **However, while Eratosthenes doubtless measured in degrees and minutes of degrees, it is unlikely that he could measure in seconds of minutes.**

Nautical measures

1 fathom = 6 feet   1 log-line = 450 feet or 75 fathoms

The fathom is believed to derive from the length of an arm-spread of lead line or anchor chain; i.e. the length of one coil. The speed of a ship used to be measured by heaving overboard the 'log' – a balk of timber fastened to a rope – and readings were noted in a 'log-book'! According to one tradition, the rope was knotted at intervals of 47' 3½" and run out for 28 seconds, whereby it was calculated that a speed of "1 knot" = 6,080 ft. per hour, because 28 x 6,080 / 3,600 = 47.29ft. The more modern log consists of a rotator with fins on it, towed astern by means of a non-twisting log line, and is used to measure distances rather than speeds.

1 knot = a speed of 1 nautical mile per hour
1 cable = 600 feet or 100 fathoms   (US 120 fathoms)
10 cables = 1 nautical mile   1 league = 3 nautical miles
1 degree = 60 (British) nautical miles = 69.091 statute miles
          60 (International) nautical miles = 69.047 statute miles

[The cable length differs from the cable attached to a ship's anchor, which is measured in *shackles* – hence the verb 'to shackle'. 1 shackle used to equal 12½ fathoms = 1/8th of 1 cable; but since revised to 15 fathoms]

So 38 statute miles = 33 nautical miles, and 15 statute miles exceed 13 nautical miles by just 160 feet, and 4,503 statute miles exceed 3,913 international nautical miles by just 1¼ inches!

The British nautical mile was fixed by the Admiralty at 6,080 ft. = 1.1515 statute miles, but was superseded by the international nautical mile of just over 6,076 ft [see EPILOGUE 9] and anyhow, in practice, was taken as 2,000 yards = 10 cable's length. The nautical mile is particularly handy, enabling the circumference of the Earth to be expressed as 21,600 miles (the number of minutes in a circle), because 21,600/60 x 69.091 (the number of miles in a degree) = 24,873 statute miles.

Whereas the geographical mile measures one minute of a degree longitude at the equator, giving a value of very nearly 6,087 ft ¼ in, the nautical mile is measured along a meridian, its length varying with latitude. [see under *Equivalents of 1 mile* below] The latitude of Athens is about 38° N and the value of the nautical mile there is just over 6,069'4". The value 6,080 ft, chosen by the Admiralty, is about right for 48°. At 50° N or S, it is 6,082 ft. (The value 1,852metres is appropriate in the region of Bordeaux). Dividing the distance from the equator to the north pole, measured along a meridian, by 5,400, the average nautical mile is nearly 6,076ft 10in, ranging from just over 6,046ft 3in at the equator to 6,107ft 6in at the pole.

1 Light-second .= 186,282.397 miles = 0.002003988805 astronomical unit

1 Light foot = the distance travelled by light in 1 nano-second = 1 sec/ 10 to the power 9

1 Astronomical unit (average between greatest and least distance of the Earth from the sun) = 92,955,807.22 miles = 499.0047835 light-seconds

1 Light-year = 5,878,499,700,000 miles (the distance light travels in a year – i.e. 1 light second x the number of seconds in 1 year)

The speed of light is known exactly because distance is defined in terms of the speed of light (i.e. 299,792,458m / sec.). But we can only know the light year with the precision with which we can measure the duration of the year, so any more precise representation would be absurd.

'Year' always means *tropical* year, as opposed to *sidereal* year, unless otherwise stated. The difference between them is explained by a phenomenon known as the precession of the equinoxes. The sidereal year is the time taken for the Earth to complete its orbit about the sun, whereas the tropical year is the time that elapses between the moment when the sun crosses the equator at the beginning of Spring until the moment when it crosses the equator at the beginning of the following Spring. The sidereal year is 365.25636 days or 31,558,149.5 seconds – and in this time the Earth rotates 366.25636 times, a day being lost by going round the sun. So the sidereal day (the time the Earth takes to rotate once) = 31,558,149.5 / 366.25636 or 86,164.1 seconds or 23 hours, 56 minutes and 4.1 seconds.

The synodic or lunar month is the period between two new moons. Its average value is 29 days, 12 hours, 44 minutes and 2.9 seconds. So there are 12.36875 lunar months in a sidereal year. In that time, the moon completes 13.36875 orbits about the Earth and thus the sidereal month, the time taken for the moon to complete one orbit, is 27 days, 7 hours, 43 minutes and 11.5 seconds.

By the way, astronomers seldom use light years. When dealing with our solar system they generally use astronomical units. Then, for the local group of galaxies they use *parsecs* and for the immense intergalactic distances they use *megaparsecs*. The 'parsec' is 648,000 / pi astronomical units or (say) 19,173,511,565,000 miles. [You can work out 'megaparsec' for yourself!]

The Julian year of 365¼ days (31,557,600 seconds) exceeds the tropical year by about 674.5 seconds. (But the Earth is slowing down, days getting longer by about 0.006 seconds a year.) The Julian year was used from 45BC until 1582 (1,626 years – there was no year 0). By then the accumulated error was over 12 days. However, the Pope decided it was only 10 days [a classic instance of the wrong use of 10 instead of 12!] and so Thursday 4 October 1582 was followed by Friday 15 October. But Britain didn't change until 1753, by which time the discrepancy was even bigger. So the cry went up: "Give us back our eleven days!" A vestige of the Julian calendar remains: the financial year ends on 5 April instead of on Lady Day (a quarter day), 25 March.

The Gregorian calendar, which we now use, is a little more complicated. The year 'N' is a leap year if 'N' is divisible by 4 but not by 100, except that, if it *is* divisible by 100 then it is a leap year only if it is divisible by 400. So the year 2000 was a leap year but 1900 wasn't!

Now from the sublime to the commonplace – miscellaneous units and measures of length

Bed sizes Compact single: 2'6" x 6'3" Popular single: 3'0" x 6'3"
Compact double: 4'0" x 6.'3" Popular double: 4'6" x 6'3"
Queen size: 5'0" x 6'6" King size: 6'0" x 6'6"

English shoe sizes – increase at 1/3rd inch intervals: size 8 = 11.333", 9 = 11.667", 10 = 12", 11 = 12.333", etc. (depending on the style of shoe)

Biblical lengths
finger	1 digit (0.91 ins)
4 digits	1 palm (3.64 ins)
3 palms	1 span (10.92 ins)
2 spans	1 cubit (21.85 ins)
6 cubits	1 reed (10.92 ft)
reed	11 feet
line	80 cubits (but can also mean a 12th part of an inch!)
mile	8 furlongs (then defined as Greek)

These are the orthodox Talmudist measures. But the 'primitive' (Zereth) measures include, interestingly enough, a span of 12.59 ins and cubit of 2 spans or 25.19 ins. That was probably the cubit that Jacob took into Egypt, where the use was made of the 'Egyptian Royal' cubit and the 'Olympic' cubit. So the Talmudist units emerged, having absorbed Babylonian influences, following the Jewish return from captivity.

Numbers used in trade

1 long (baker's) dozen	= 13	1 gross	=	144
1 long gross	= 156	1 score	=	20
1 common hundred	= 100	1 long or great hundred	=	120

Deriving from the human body

1 digit	=	width of middle finger (¾ inch)
1 inch	=	………..thumb across knuckle
1 nail	=	2¼ inches (from middle joint to tip of middle finger)
1 ell	=	20 nails = 45 inches (from one shoulder across chest to finger-tip of outstretched arm – used for measuring cloth – try it!)
1 palm	=	3 inches
1 hand	=	4 inches (width of palm plus adjacent thumb joint)
1 shaftment	=	9 digits (hand plus length of extended thumb – 6.00/ 6.55 ins)
1 span	=	9 inches (tips of outstretched thumb to little finger)
1 cubit	=	18 inches (from elbow to outstretched finger tips
1 fathom	=	6 feet (from tip to tip of fingers of outstretched arms)

Equivalents of 1 inch   25,400.05 microns*   4,000 silversmiths' points
1,000 mils (mils are used for gauging wire and firearm bores)
67.387 UK or 72.27 US printer's points (respectively 0.1484" and 0.1384")
64 *ounces* (to measure thickness of shoe leathers: also a term in Troy weights)
23.24635 agates (also used in printing)   48 hairsbreadths
12 *lines* (used in cloth measure and in printing)   3 barleycorns
1.3333 digits (4/3)   0.3333 palm (1/3)   0.25 hand (1/4)
0.08333 foot (1/12)   0.055556 cubit (1/18)   0.027778 yard (1/36)
0.005051 rod (1/198)   * microns relate to the 1893 American inch

Several of these equivalents are not accurate because expressing fractions by decimals often requires use of 'repeaters' – a major defect of the metric system.

Equivalents of 1 foot (see also Appendix 1)   12,000 mils   144 lines
4 palms   3 hands   0.66667 cubit (2/3)   0.3333 yard (1/3)
0.060606 rod (2/33)   0.001515 furlong (1/660)

Equivalents of 1 mile   5,280 feet   3,520 cubits   1,760 yards   320 rods
8 furlongs   0.868978 meridian mile   0.868421 British nautical mile
0.867419 geographical mile

The *meridian* mile was established by international agreement in 1954 to represent 1 minute (1/60th of 1 degree) of the Earth's meridian, but has been generally replaced by the nautical mile, as also has the *geographic* mile, formerly known as the *admiralty* mile. In 1929 the International Hydrographic Bureau proposed a length of 6,076.097 feet (Why, for goodness' sake?) but compare with NAUTICAL MEASURES.

The degree (1/360th of the Earth's equatorial circumference) is taken to equal 69.1707 statute miles; but 1° of latitude equals 68.708 miles at the equator and 69.403 miles at the poles, as determined by the International Astronomical Union Ellipsoid of 1964, while 1° of longitude equals 69.1707 miles at the equator diminishing to zero at the poles. [see NAUTICAL MEASURES]

Surveyors' measures

1 link = 7.92 inches
1 pole = 25 links = 198 inches = 16½ feet
1 chain = 100 links (792 inches) = 4 poles = 66 feet
1 furlong = 10 chains = 1,000 links = 40 poles = 660 feet
1 mile = 80 chains

(The circumference of a circle of 7 yards diameter is very close to 1 chain.)

Decimal conversions:

to change *feet* into *miles (nautical)* multiply by 0.0001645
        (land)   0.0001894
   feet per second into *m.p.h.*   0.6818
   miles (nautical) into *miles (land)*   1.1515
   miles (land) into *miles (nautical)*   0.8684
   m.p.h. into *feet per minute*   88

CIRCULAR MEASURES

1 minute = 60 seconds 1 degree = 60 minutes 1 sign (rare) = 30 degrees
1 sextant = 60 degrees   1 quadrant = 90 degrees   1 grade = 0.01 quadrant
1 circle = 360 degrees = 32 points = 4 quadrants = 400 grades =
2 x 22/7 radians   1 radian = 57.2958 degrees:
1 circle = 2 x *pi* x 1 radian = 2 x 3.14159 x 57.2958 = 360 degrees

WEIGHT AND MASS

Edited from Colin R Chapman's 'How Heavy, How Much and How Long' (Lochin Publishing Society 1995):

"In many cases we should be identifying mass, not weight. Strictly, the weight of an item is the force exerted on it by gravity, while mass is the amount of matter in that item. If, for example, the absolute weight of an item is measured by a spring-balance at the bottom of a mine and then on the top of a mountain, it will weigh less on the mountain top as the gravitational attraction there is less, even though its mass has not altered. On the moon, where the gravitational force is a sixth of that on earth, the item will weigh six times lighter and a spring-balance will indicate this.

In practice, for goods being weighed in a particular market on a beam balance or a steelyard [a type of balance having its fulcrum off-centre, enabling heavy goods to be weighed with relatively small weights: e.g. a 7 lb weight at 24 in. from the fulcrum counter-balancing a 28 lb item at 6 in. on the other side of the fulcrum], the difference between mass and weight is of no consequence; two items with equal masses having equal weights under identical conditions. So the veracity of any iron and brass weights is important and therefore national and local standards, which were the subject of a number of Acts of Parliament, were used to check the weights used by tradesmen."

Avoirdupois Weights (although more properly *'avoir de pois'*)

1 ounce (oz) = 16 drams   1 pound (lb) = 16 ounces   1 stone = 14 lb
2 stone = 1 quarter   1 hundredweight (cwt) = 8 stone = 112 pounds
1 ton = 20 hundredweight = 160 stones = 2,240 pounds
1 short ton = 2,000 pounds

'Pound' derives from the Latin *pondus* meaning weight, and the abbreviation 'lb' from the Latin *libra* meaning pound. 'Ounce', like 'inch', comes from *uncia* meaning a twelfth part.

King Edward I had a 'merchant's pound' of 15 x 450 grains = 6,750 grains. Edward III's *haber de pois* weights (there's a set in Winchester Museum) were 16 Florentine ounces x 437 grains (compare the Roman ounce of 438 grains) = 6,992 grains, which Elizabeth I rounded up to 7,000.

So the Avoirdupois pound, in common use by the 14th century, was based on 16 ounces, equating to 7,000 grains; the avoirdupois ounce equalling (as it still does) 437½ grains. The ease of 16 for division, especially in the wool trade, led to this system s becoming standard. Specific containers — sacks, barrels, casks, pockets, tubs, chests, baskets, drums, etc — became identifie as standard measures for specific commodities. But wool was always a special case, because of the wealth its trade created and the tax revenue levied on it, based on weight measured in *trones* (scales) using 7 and 14 lb lead or bronze weights. The stone of 14 lb is still widely used in the wool, fish and vegetable trades, as well as for weighing people.

Troy Weights (formerly used for precious metals and gems)

24 grains = 1 pennyweight (dwt)   **20 dwts = 1 ounce**   **12 oz = 1 lb**

The Troy pound equals 144/175 of the pound avoirdupois but the Troy ounce equals 192/175 of the ounce avoirdupois. An ancient English measure of weight (though named after the French city of Troyes), the Troy pound equated to the monetary pound sterling that contained 240 pennies – a penny originally containing a pennyworth of almost pure silver (37/40 silver with 3/40 copper) which had the same weight as 24 grains of barleycorn. An Act of 1853 declared the Troy ounce of 120 *carats* to be the standard for the sale of bullion.

The Troy pound was abolished in 1878, but the Troy ounce still survived (as 1.097oz avoirdupois) in the Weights and Measures Act of 1985.

Apothecaries Weights (formerly used by pharmacies)

1 scruple = 20 grains   1 drachm = 3 scruples
1 ounce = 8 drachms   1 pound = 12 ounces

The Apothecaries pound and ounce corresponded to the Troy pound and ounce, being equal in weight respectively to 5,760 and 480 grains of barleycorn. But the Apothecaries system excluded pennyweights and properly excluded pounds too; dividing instead into drachms and scruples. It was abolished in Britain in 1971. Troy and Apothecaries Weights remain full of interest, however, and references to them are frequently found, as likewise to ancient maritime and mercantile measures, which were universal and essential in their day, such as in this typical shipping report from 1821: *"Arrived Greenock, 15 January, 'The Royal Charlotte' from Lisbon, with 183 chests and 334 half-chests oranges, 4 pipes and 4 hogsheads white wine, ¼ cask port wine, 144 quintals of salt and 10 dozen mats."* A quintal was a weight of 100 lb, often used for fish, while a mat(t) was a weight of 80 lb, used for sugar and spices.

Maundy pennies of sterling silver are still minted but they weigh only seven & three elevenths grains, comprising six & eight elevenths grains of silver plus six elevenths grains of base metal — because sterling silver is 37 / 40ths pure silver.

MEASURES OF AREA IN EVERYDAY USE

1 square foot = 144 square inches   1 square yard = 9 square feet
1 square rod/ pole/perch = 30¼ square yards
1 rood = 1 furlong x 1 rod = 40 sq. rods = 1,210 sq. yards = ¼ acre

1 acre = 4,840 square yards = 1 furlong x 1 chain = 1 furlong x 4 rods/ poles/ perches = 4 rods x 40 rods = 1 chain x 10 chains

1 square mile = 640 acres (i.e. 1,760 x 1,760 = 4,840 x 640)

The acre was originally the extent of land that a horse or ox could plough in one day. *('linacre' may derive from 'field of flax', or is it an abbreviation of 'linear acre'?).* It was 40 rods or a 'furrow long' (furlong) by a width divided into 72 furrows, 11 inches apart: i.e. 792 inches, 100 links, 22 yards, 4 rods or 1 chain.

Gunter's square land measure

1 square link = 62.7264 square in.   1 sq. rod/ pole = 625 sq. links
1 square chain = 16 square poles or 10,000 square links
1 acre = 10 square chains

Builders' square measures

1 square of flooring = 100 sq.ft.   1 rod of brickwork = 272 ½ sq.ft.
1 bay of slating = 500 square feet   1 'yard' of land = 30 acres
1 'hide' of land = 120 acres (the 'long hundred')

MEASURES OF VOLUME AND CUBIC CAPACITY

**1 cubic foot = 1,728 cubic inches   1 cubic yard = 27 cubic feet
1 cu.ft. of water = 6.25 gallons weighing 1,000 ounces (c. 6.25 x 8 x 20)
1 standard gallon = 277.274 cubic inches
1 standard bushel = 8 gallons = 2,218.19 cubic inches (both obsolete)**

1 cubic foot is very nearly 6.228835459 gallons and 1 cubic foot of water at a temperature of 62°F weighs very nearly 997.68 ounces. [see APPENDIX V]

The standard gallon as defined in the 1824 Act; but revised to 277.421 cubic inches in 1932: but since 1878 a standard Imperial gallon has been defined as the capacity of 10 *avoirdupois* pounds of distilled water at 62°F and 30 inches of mercury barometric pressure. [see APPENDIX V]

1 Acre-Inch = 3,630 cubic feet   **1 Acre-Foot** = 43,560 cubic feet
(These measures are used mainly in relation to irrigation and flood control)

FLUID MEASURES

1 pint = 4 gills = 8 tots /noggins = 20 fl.oz
1 gallon = 4 quarts = 8 pints
1 bushel = 4 pecks = 8 gallons

1 quart = 2 pints
1 peck = 2 gallons

'Pint' appears to derive from the Latin *pingo* meaning 'I paint', from the mark that used to be painted on a vessel to indicate the pint measure. Bushels and pecks were customary measures, particularly for dry goods such as grain, salt and flour, even fish and coal, while less familiar measures were also employed. Barrels and hogsheads, with pints and gallons, were used as measures of dry capacity as well as indicating volumes of fluids – although the absolute size of a barrel depended very much on the nature of its contents. Like the barrel, the gallon had an erratic pedigree, being defined in 1290 by its capacity for 8 pounds of wheat, then in 1706 as containing 231 cubic inches – which remains the USA standard. (The Pilgrim Fathers sailed in 1620 but the mass migrations took place much later.) The royal decree that led to the creation of dry and liquid measures had been King Henry VII's in the late 15th century, declaring that "eight pounds do make a gallon of wine and 8 gallons do make a bushel". In the USA they still do.

1 US liquid ounce = 1.04084 fl.oz = 1.80469 cu.in.
1 UK fl.oz. = 1.73387 cu.in.
1 cu.ft. = 6.22884 UK gallons = 7.48052 US gallons = 0.178108 US barrels
1 US gallon = 0.832674 UK gallons = 0.0238095 US barrels = 0.133681 cu.ft.
1 UK gallon = 0.028594 US barrels = 277.419 cu.in. = 0.160544 cu.ft.
1 US barrel = 42 US gallons = 34.9723 imperial gallons = 5.61458 cu.ft.
1 US pint/gallon = 0.832674 UK pint/gallon
1 UK pint/gallon = 1.20095 US pint/gallon

In the Kitchen

2 saltspoons	=	1 teaspoon	2 teaspoons	=	1 dessertspoon
2 dessertspoons	=	1 tablespoon	4 tablespoons	=	1 teacup
2 teacups	=	1 breakfastcup	2 breakfastcups	=	1 pint
1 pint	=	20 fl.oz	1 breakfastcup	=	10 fl.oz
1 teacup	=	5 fl.oz	1 tablespoon	=	1 1/4 fl.oz
1 dessertspoon	=	5/8ths fl.oz	1 teaspoon	=	1/3rd fl.oz
1 saltspoon	=	1/6th fl.oz			

British measures

1 teaspoon	= 5/16ths fl.oz
1 tablespoon	= 1 1/4
1/4 pint	= 5
1/2 pint	=10
3/4 pint	=15
1 pint	=20 fl.oz

US measures

6 teaspoons	=	1 liquid oz
2 tablespoon	=	1 liquid oz
1 cup	=	8 liquid oz
1 US pint	=	16 liquid oz

Wine and champagne   1 bottle = 1/6th of 1 gallon   1 magnum = 2 bottles
    1 Jeroboam = 4 bottles   1 Rehoboam = 6 bottles (1 gallon)
    1 Methuselah = 8 bottles   1 Salmanazar = 12 bottles
    1 Balthazar = 16 bottles   1 Nebuchadnezzar = 20 bottles

Beer and ale   1 jug of ale = 1 pint   1 tankard = 1 quart
1 flagon or pitcher = 2 quarts = 1/2 gallon   1 firkin = 9 gallons
1 kilderkin = 2 firkins   1 barrel = 4 firkins
1 hogshead = 3 kilderkins (54 gallons)   1 bottle = 1 1/3 pints = 75.7682cls
1 puncheon = 2 barrels (72 gallons)

Originally, the British gallon weighed 8 pounds, whether liquid or corn (for dry goods); the pint bearing a 1:1 relationship to the pound. But when the Imperial system was regularized under George IV, a larger Tudor corn gallon holding 10 pounds of water was adopted. However, because the true liquid range of measures from the Queen Anne era had already been taken to the American colonies, they became standard for the USA. [see INTRODUCTORY ARTICLES pages 3-4 and FLUID MEASURES]

Apothecaries' fluid measures

1 fluid scruple = 20 minims   1 drachm = 3 scruples
1 fluid ounce = 8 fluid drachms = 24 fluid scruples
1 pint = 20 fluid ounces   1 corbyn (rare) = 40 fluid ounces = 2 pints

Medicine measures   (Note: these differ from *In the Kitchen*!)

1 teaspoon = 1 fluid drachm = 60 drops or minims
1 dessertspoon = 2 fluid drachms   1 tablespoon = 4 fluid drachms
1 wineglass = 2 fluid ounces   1 teacup = 3 fluid ounces

PAPER SIZES (inches) – little used now but of antiquarian and aesthetic value

NAME	WRITING & DRAWING	PRINTING	WRAPPING
Emperor	72 x 48		
Double Quad Crown		60 x 40	
Quad Imperial			58 x 45
Double Nicanee			56 x 45
Quad Royal			50 x 40
Antiquarian	53 x 31		
Casing			46 x 36
Saddleback			45 x 36
Quad Demy		45 x 35	
Double Imperial			45 x 29
Grand Eagle	42 x 28¾		
Quad Crown		40 x 30	
Double Elephant	40 x 26¾		
Double Royal		40 x 25	
Colombier	34½ x 23½		
Atlas	34 x 26		
Double Large Post	33 x 21	33 x 21	
Six Pound Grocers			32 x 22
Double Four Pound			31 x 21
Double Demy	31 x 20	35 x 22½	
Double Post	30½ x 19	31½ x 19½	
Imperial	30 x 22	30 x 22	
Double Crown		30 x 20	
Imperial Cap			29 x 22
Elephant	28 x 23	28 x 33	34 x 24
Super Royal	27 x 19	27½ x 20½	
Double Foolscap	26½ x 16½	27 x 17	
Cartridge	26 x 21		
Haven Cap			26 x 21
Four Pound Grocers			26 x 20
Royal	24 x 19	25 x 20	
Sheet & half Foolscap	24½ x 13½		

Bag Cap		24 x 19½
Sheet & half Post		23½ x 19½
Medium	22 x 17½	23 x 18
Sheet & third Foolscap	22 x 13½	
Kent Cap		21 x 18
Large Post	21 x 16½	
Copy or Draft	20 x 16	20 x 16½
Demy	20 x 15½	22½ x 17½
Music Demy		20 x 15½
Crown		20 x 15
Post	19 x 15¼	19¼ x 15½
Pinched Post	18½ x 14¾	
Foolscap	17 x 13½	17 x 13½
Brief	16½ x 13¼	
Pott	15 x 12½	

[Oliver Cromwell, when asked whether the water-mark of the king's head should remain on state papers, replied that "the old fool's head" could continue to be used. But 'head' was thought indelicate as it had been chopped off; hence 'foolscap']

Paper folding into leaves & pages

NAME	Abbreviation	Folded into leaves	Folded into pages
Folio	Fo	2	4
Quarto	4to	4	8
Octavo	8vo	8	16
Duodecimo	12mo	12	24
Sextodecimo	16mo	16	32
Octodecimo	18mo	18	36
Vicesimo	20mo	20	40
Vicesimoquarto	24mo	24	48
Duoettricesimo	32mo	32	64

Sizes of bound books

NAME　　　　　　　Size (inches)　　　　Abbreviation

(Sizes given in height x width, but books whose width exceeds their height are marked 'ob':   e.g. 'ob Crown Folio (obCfol)' is 10" high x 15" wide)

Name	Size (inches)	Abbreviation
Imperial Folio	22 x 15	Impfol
Super Royal Folio	20 x 13½	suRfol
Royal Folio	20 x 12½	Rfol
Medium Folio	18 x 11½	Mfol
Demy Folio	17½ x 11¼	Dfol
Crown Folio	15 x 10	fol
Post Folio	15¼ x 9½	Post fol
Foolscap Folio	13½ x 8½	Ffol
Pott Folio	12½ x 7¾	Pottfol
Imperial Quarto (4to)	15 x 11	Imp4
Super Royal 4to	13½ x 10	suR4
Royal 4to	12½ x /10	R4
Medium 4to	11½ x 9	M4
Demy 4to	11¼ x 8¾	D4
Crown 4to	10 x 7½	C4
Post 4to	9½ x 7⅝	Post4
Foolscap 4to	8½ x 6¾	F4
Pott 4to	7¾ x 6¼	Pott4
Imperial Octavo (8vo)	11 x 7½	Imp8
Super Royal 8vo	10 x 6¾	suR8
Royal 8vo	10 x 6¼	R8
Medium 8vo	9 x 5¾	M8
Demy 8vo	8¾ x 5⅝	D8
Large Crown 8vo	8 x 5¼	
Crown 8vo	7½ x 5	C8
Post 8vo	7⅝ x 4⅞	Post8
Foolscap	6¾ x 4¼	F8
Pott 8vo	6¼ x 3⅞	Pott8
Demy 16mo	5⅝ x 4⅜	D16
Demy 18mo	5¾ x 3¾	D18
Music	14 x 10¼	

MUSIC

**1 Semibreve = 2 minims, 4 crochets, 8 quavers,
16 semiquavers, 32 demisemiquavers, 64 hemidemisemiquavers.**

This sequence of numbers is compatible with the imperial system of measures but alien to the metric system. Likewise, an *Octave* is a tone on the eighth degree from a given tone counted as the first. So one octave note is exactly double the frequency of the same note in an octave below it, and half that of the one above it. For example, the note A = 440 vibrations a second and A above AQ = 880 vibrations per second.

The range can be seen and heard on a piano where eight successive white keys equal an octave. A *semiditone* or *hemiditone* is a minor third, especially in Greek music, and a *sextuple* is a unit characterized by having six beats to the measure. (The ratio of minor tone to major tone is 80:81 and is called the 'comma of Pythagoras'; but, with a 'well-tempered' keyboard – as required to perform Bach's 48 preludes and fugues which were written in all the keys imaginable – the distinction between major and minor tones disappears.) It was Pythagoras who discovered the basic relation between mathematics and musical harmony, from the fact that the vibration of a single stretched string produces a ground note, the notes that sound harmonious being produced by dividing the string into an exact number of parts – into exactly two parts or exactly three or four parts and so on.

If the still point on the string (the *node*) does not come at one of those exact points, the sound is discordant. As the node is shifted along the string, recognizably harmonious notes are sounded as the prescribed points are passed. From the ground note of the whole string, then with the node at the midpoint the note is one octave higher; at a point one third of the way along the note is a fifth higher; and so on.

So it was Pythagoras who found that the chords that sounded most pleasing to the Western ear correspond to exact divisions of the string by whole numbers. His followers believed that the agreement between nature and number was so significant that even the movement of celestial bodies could be calculated in relation to musical intervals. [Anne Macaulay showed – see **Megalithic measures** – that the geometry of the great Stone Circles of 5,000 years and more ago does indeed correspond to this musical mathematics.]

There are no more 5s or 10s in the measurement of music than in the measurement of time. Halving and doubling is how the human brain works. Favourite popular time signatures are 2 /4, 3 /4, 4 /4 and 6 /8. The essence of the *binary* system of numbering, which governs computer arithmetic – the basis of modern technology – is counting in twos, not tens. The value of any binary number increases by power of 2 (i.e. doubles) with each move from right to left (1, 2, 4, 8, 16, etc.) and *vice versa*.

A Note on Organs (with thanks to Harry Coles): Whether an organ is destined for Oswestry, Oslo or Osaka, its No. 1 Diapason on the Great (Principal) will speak at 8 ft pitch. The imperial foot has ruled since before the monk Wulstan (d. 963) played on Winchester Cathedral's organ. If one draws a stop labelled 8 ft and starts playing on its appropriate manual, its pitch – the number of vibrations per second for any given note (fast ones producing high pitch, slow ones low) – is the same as when playing the piano. Now by a quirk of natural physics, the note C (the third white note up from the bottom on a standard keyboard) is produced by an ordinary wood or metal organ pipe from its mouth to its top, just 16 ft in length. Twelve pipes on to the next C, and they have tapered down to 8 ft; another twelve and that pipe is only 4 ft, and so on. This is yet another illustration of the natural progression or regression of numbers, which is wholly at odds with decimal arithmetic. Some pipes for very high notes are only a few inches long. A large organ such as in Liverpool's modern Anglican Cathedral contains thousands of pipes, all fashioned to precise specifications. In contrast, a Victorian 'Father Willis' church organ has stops of 16, 8, 4 and 2 ft pitches – and so will any similar organ, whether in Tokyo, Timbuktu or Tewkesbury!

SOUND

The speed of sound in air varies with temperature. It's about 1,125 ft per second at 68°F but just under 1,087 ft/sec at 32°F. The normal standard is 1,100 ft/sec which is exactly 1 mile in 4.8 seconds. Sound travels much faster in water, at about 4,760 ft per second, or 1 mile in just 1.1 seconds.

PRESSURE

Different scales are used to measure. A barometer balances atmospheric pressure against the pressure exerted by a column of mercury whose height is measured in inches. Normal atmospheric pressure at sea level is about 30in. of mercury, which is adopted as the British standard. The standard in the metric system is defined as 101.325 kilopascals which comes to very nearly 760mm of mercury. But 30in. = 762mm; whereas 101.325 kilopascals corresponds to a pressure of very nearly 2,116.22 lb/sq.ft or 14.6959 lb/sq.in. *Tyre* pressures are measured in lb per sq.in.; but when we say that the air pressure in a tyre is 25 lb per sq.in., we mean that this is the amount by which that pressure *exceeds* atmospheric pressure. *Blood* pressure is measured in the UK in mm. of mercury (whereas the French use *cm*!).

Now, how are these three scales related? Well, 1 lb per sq.in. is *very* nearly 51.7149 millimetres or 2.03602 inches of mercury. (Now you know!)

TEMPERATURE

The Fahrenheit scale, named after Gabriel Daniel Fahrenheit (1686-1736), a German physicist. Initially, he took 100° to represent the temperature of his own body, which he later discovered to be about 98.3-98.4°. There are two theories as to how he chose **0** on his scale: either it was on an extremely cold day in Danzig (Gdansk) in 1709 or it was the temperature of a mixture of salt and ice. The modern keys to the Fahrenheit scale are: absolute zero, which is *minus* 459.67°, and the *triple point* of water – the temperature at which ice, water and water vapour can co-exist – which is 32.018°F.

The temperature of the triple point of water on the Kelvin scale is 273.16°; this number being chosen so that the size of the kelvin should match the Celcius degree. The Rankine scale does for Fahrenheit what the Kelvin scale does for Celsius. The Kelvin scale, named after William Thomson, Lord Kelvin (1824-1907), is based on thermo-dynamic principles. Fahrenheit, Rankine and Celsius are each given in degrees, but kelvin is just kelvin.

The following table shows a sample of temperatures, on these four scales, relating to: absolute zero, freezing point of water, body temperature, and boiling point of water at standard pressure.

° kelvin	Celsius	Fahrenheit	Rankine
0	- 273.15	- 459.67	0
273.15	0	32	491.67
310.05	36.9	98.42	558.09
373.15	100	212	671.67

Incidentally, it is a vital principle of the metric system that only one unit is authorized for each type of magnitude. Therefore, since the kelvin is the metric unit of temperature, the degree Celsius does not belong to the metric system.

FORCE

The *poundal* is the force which, applied to a particle of mass 1 pound, gives it an acceleration of 1 ft. per second per second. If a particle starts from rest, accelerating at that rate, it is moving at a speed of 1 ft per second after 1 second 2 ft per second after 2 seconds, 3 ft per second after 3 seconds, and so on. A particle of mass 1 lb falling freely (i.e. acted upon only by its own weight) has an acceleration of about 32.19 ft/sec/sec; so a particle of mass 1 pound has a weight of about 32.19 *poundals*. (The unit of force in the metric system is defined in the same way. The newton is the force which, applied to a particle of mass 1 kilogram, gives it an acceleration of 1 metre/sec/sec.)

The force equal to the weight of a particle whose mass is 1 lb is called the lb weight or pound force. In practice, the word weight or force is usually omitted, so confusion often arises. In the British system we thus have two basic units of force: the poundal and the pound weight. One is based on *inertia* and the other on *gravity*. While the lb weight is used for everyday purposes, the poundal is preferred for scientific work – partly because weight varies slightly with location on the planet.

The approximate weight of a particle of mass 1 lb at different latitudes is shown in the following table, which demonstrates the relevance of references to latitude in the section on work and energy.

latitude (degrees)	weight (poundals)
0	32.0877
15	32.099
30	32.1301
45	32.1726
60	32.2152
75	32.2464
90	32.2578

The standard (legal) formula for converting pounds weight to poundals is: 6,096 lb weight = 196,133 poundals. This gives the weight of a particle of mass 1 lb at a latitude of about $45\frac{1}{2}°$.

WORK and ENERGY

To raise a particle of mass 1 lb by 1 ft, 1 ft.lb weight (or 1 foot pound) of energy is needed. Likewise, if something is moved 1 ft against a resisting force of 1 poundal, then 1 foot poundal of work is done.

Following the discovery that heat is a form of energy, the English physicist James Prescott Joule * (1818-1889) – working, of course, in English units – found that, in the latitude of Manchester, 1 British Thermal Unit equalled 773.64 ft.lb. The legal conversion formula is: 7.4726673 BTU = 5,815 ft lb. 1 BTU represents the amount of heat required to raise the temperature of 1 lb of water at its greatest density (occurring at about 39°F) by 1°F. Thus, the heat required to raise the temperature of 1 pint of water ($1\frac{1}{4}$ lb) from 52°F to boiling point – a rise of 160°– is 160 x $1\frac{1}{4}$ = 200 BTU. The *therm* = 100,000 BTU.

[*The *joule* is the *Systeme International* unit of work and energy, replacing the calorie – which is *not* a metric unit. It is defined as the work done (energy transferred) by a force of one newton acting over one metre. The calorie is defined as 4.1868 joules and the BTU as very nearly 1.05506 kilojoules.]

POWER

Power is the rate at which energy is converted from one form into another. In the metric system, the unit of power is the ***watt***, named after the Scottish engineer James Watt (1736-1819). 1 watt is a rate of working of 1 joule per second: e.g. a 1 kilowatt electric fire turns electrical energy into heat at the rate of 1,000 joules per second. Before the advent of the steam engine, motive power was usually supplied by horses. So, when James Watt sold his steam engines, he had to tell his customers how many horses they would replace. He found that a horse would walk comfortably at 2½ m.p.h. while pulling against a resistance of 100lb. Now, 2½ m.p.h. = 13,200 ft. in 3,600 seconds = 3.66 ft/sec. So the horse was working at a rate of 366.66 ft.lb/sec. Watt didn't want to overstate his case, so he added a half again to this figure, giving 550 ft.lb/sec. That, then, is the British horse-power.

The Continentals decided on 75 metre kilograms/sec. for their *'cheval-vapeur'*. So the British HP = 745.69987158227022 watts (the official government figure and the longest in this book!), whereas the European HP = 735.49875 watts.

A machine HP is usually reckoned to equal the power of 4.4 horses, one horse being assumed to possess the strength of 5 men; so 1 HP notionally matches the strength of 22 men. But that was entirely fanciful. For if an 11 stone man pedals a 6 lb bicycle up a 1 in 20 hill at 10 mph he is doing work at the rate of more than $((11 \times 14) + 6) \times 10 \times 22 / 15 \times 20$ ft lbs / sec = 117.33 ft lbs / sec or 0.2133 HP – which is nearly 5 times the rate of working suggested by 22 men being equivalent to 1 HP!

Horse-power (as in motor-cars) is determined either as *indicated* HP, which is the power developed in the engine cylinders as calculated from (a) the average pressure of the working fluid (b) the piston area (c) the stroke and (d) the number of working strokes /minute, or as *brake* HP, which is the actual power of the engine calculated from (a) the force exerted on a friction brake or absorption dynamometer acting on a fly-wheel or brake-wheel rim (b) the effective radius of this force and (c) the speed of the fly-wheel or brake-wheel.

Imperial and metric units are often used together. Thus, it takes about 1.023 megajoules to convert 1 lb of water to 1 lb of steam at 212°F. For 34½lb, 35.29 megajoules are needed, and if this is done in 3,600 seconds, the rate of working is 9.802 kilowatts or 13.15 horse-power.

SPEED

One of the many beauties of the imperial system is the compatibility between its units of distance and the measurement of time, because of the duodecimal factor common to both: e.g. 1 m.p.h. = 1,056 in./minute = 88 ft/minute; and 1 mile per day (24 hours) = 220 ft/hr., etc.

A Mach number, named after Ernst Mach (1838-1916) the Austrian physicist, indicates the speed ratio of an object to the speed of sound – about 1,087 feet per second through air at sea level at 32°F. So Mach 1 equals the speed of sound; while hypersonic speed means Mach 5 or faster.

[We now slow down considerably!]

AN ANTHOLOGY

Then down with every metric scheme
Taught by the foreign school;
We'll worship still our father's God
And keep our father's rule –
A perfect inch, a perfect pint,
The honest British pound,
Shall hold their place upon the Earth
Till time's last trump shall sound   (from a popular 19th century song).

An old-timer's account of the causes of an Australian rural recession in the 1980s (anon)

"It all started in the sixties when they changed pounds to dollars: that doubled me overdraft. Then they brought in kilograms instead of pounds: me wool-clip dropped by half.
After that they changed rain to millimetres: and we haven't had an inch of rain since. If that wasn't enough, they brought in Celsius, so we never get above forty degrees – no wonder the wheat don't grow.
Next came hectares instead of acres; so I only had half the land I used to own. By this time I'd had enough, and decided to sell out. I put the property into the agent's hands, but then they changed miles into kilometres, so now I'm twice as far out of town and nobody wants to buy the b----- place!"

An ounce is to dangle from your finger
A pound is to hold in your hand
A stone is to carry under your arm
A hundredweight over your shoulder
And a ton a horse pulls in a cart   *(traditional)*

from the Bible

Isaiah xl, 12 & 21-22: Who hath measured the waters in the hollow of his hand, and meted out heaven with the span, and comprehended the dust of the earth in a measure, and weighed the mountains in scales and the hills in a balance? 'Have ye not known, have ye not heard, hath it not been told to you from the beginning, have ye not understood from the foundations of the earth? Is it not he that sitteth on the circle of the earth...'

Daniel v, 25: Mene, Mene, Tekel, Upharsin. This is the interpretation of the thing: Mene – God hath numbered thy kingdom, and finished it; Tekel – Thou art weighed in the balances and art found wanting.

Zechariah ii, 1-2: I lifted up my eyes again, and looked, and behold a man with a measuring line in his hand. Then said I, Whither goest thou? And he said unto me, To measure Jerusalem, to see what is the breadth thereof, and what is the length thereof.

Job xxxiix, 5-7: Who hath laid the measures thereof, if thou knowest? Or who hath stretched the line upon it? Whereupon are the foundations thereof fastened?

Ezekiel xl, 5 & xliii, 13-15: And behold a wall on the outside of the house round about, and in the man's hand a measuring reed of six cubits long by the cubit and an hand breadth: so he measured the breadth of the building, one reed; and the height, one reed. And these are the measures of the altar after the cubits: The cubit is a cubit and a hand breadth; even the bottom shall be a cubit, and the breadth a cubit, and the border thereof by the edge thereof round about shall be a span: and this shall be the higher place of the altar. And from the bottom upon the ground even to the lower settle shall be two cubits, and the breadth one cubit; and from the lesser settle even to the greater settle shall be four cubits, and the breadth one cubit. So the altar shall be four cubits; and from the altar and upward shall be four horns. *[Yes, most obscure!]*

Proverbs: xx. 10: Divers weights and divers measures…

Matthew: v. 41: Whosoever shall compel thee to go a mile, go with him twain.

others (chronologically)

Hesiod (9th century BC): Hew a mortar three feet in diameter, and a pestle three cubits, and an axletree seven feet…and hew a wheel of three spans for the plough-carriage of ten palms.

Herodotus (BC484-424): All men who are short of land measure it by fathoms; but those who are less short of it, by stadia; and those who have much, by parasangs; and such as have a very great extent, by schoinoi. Now a parasang is equal to 30 stadia, and each schoinos, which is an Egyptian measure, is equal to 60 stadia.

Plato (BC c.428-347): The most important and first study is of numbers and the greatest of their influence on the nature of reality…Every diagram and system of number and every combination of harmony and the agreement of the revolution of the stars must be made manifest as one in all to him who learns in the proper way, and will be made manifest if a person learns aright by keeping his eyes on unity…There is a natural bond linking all things.

Cicero (BC106-43): Not to know what happened before we were born is to remain perpetually a child. For what is the worth of a human life unless it is woven into the life of our ancestors by the records of history?

Flavius Josephus (AD c.37-100): The author of weights and measures, an innovation that changed a world of innocent and noble simplicity, in which people had hitherto lived without such systems, into one forever filled with dishonesty.

Leon Battista Alberti (1440): We shall ever give ground to honour. It will stand to us like a public accountant; just, practical, and prudent in measuring, weighing, considering, evaluating, and assessing everything we do, achieve, think and desire.

St Isadore of Seville (c.1600): Take away number in all things and all things perish. Take calculation from the world and all is enveloped in dark ignorance; nor can he who does not know the way to reckon be distinguished from the rest of the animals.

Milton (1608-74): In his hand / He took the golden Compasses, prepar'd / In God's eternal store, to circumscribe / This universe, and all created things: / One foot he center'd, and the other turn'd / Round through the vast profunditie obscure / And said, thus far extend / Thus far thy bounds / This be thy just Circumference, O World. *(Paradise Lost, Book VI)*

Napoleon Buonaparte (1769-1821): The scientists adopted the decimal system on the basis of the metre as a unit. Nothing is more contrary to the organisation of the mind, memory and imagination. The new system will be a stumbling block and source of difficulties for generations to come. It is just tormenting the people with trivia.

Robert Southey (1774-1843): An ounce of love is worth a pound of knowledge. *(Life of Wesley: ch.16)*

Lord Kelvin (1891): I often say, when you can measure what you are speaking about and express it in numbers, you know something about it; but when you cannot measure it, when you cannot express it in numbers, your knowledge is of a meagre and unsatisfactory kind.

Lewis Carroll (1832-98): Rule forty-two. All persons more than a mile high to leave the Court. *(Alice in Wonderland: ch.12)*

W H Auden (1907-73): And still they come, new from those nations to whom the study of that which can be weighed and measured is a consuming love.

MEGALITHIC MEASURES

The origins of Imperial measures derive from mankind's earliest knowledge of astronomy and geometry, of music and the measurement of time. But in pre-literate history, before the discovery of any means of recording and handing such knowledge down to succeeding generations, it is impossible to trace the rudiments of these sciences farther back than the late Neolithic period of around 4,000BC. This was otherwise known as the Megalithic period, because it was distinguished chiefly by the creation of the world's earliest structures, in the shape of the great stone rings, which were erected throughout what are now the British Isles and NW France.

The remains of some 900 are still standing, of which some 200 of the most prominent were meticulously surveyed by Alexander Thom (Professor of Engineering Science at Oxford University from 1945 to 1961) and his son A S Thom (senior lecturer in the Department of Aeronautics and Fluid Mechanics at Glasgow University). Their analyses were published in three major books (OUP) from 1967 to 1978. They proved the astronomical purposes for which several of these colossal monuments were built and also that they were all planned and engineered on the basis of two key units of length: the ***Megalithic Yard*** **equal to 2.72 ft and the** *Megalithic Rod* **of 6.8 ft; so that 5 MY = 2 MR**. They identified, furthermore, many distinct types of geometric layout.

Thoms' work has been vastly extended and elaborated by Anne Macaulay (1924-1998), the results of whose research, throughout the last thirty years of her life, are being published in a book next year (2001). She showed that these Megalithic people on the Atlantic seaboard virtually invented mathematics (whether even more ancient civilizations in China had done so independently is another question), for they must have devised the Fibonacci series of numbers almost 5,000 years before the birth of Leonardo of Pisa (c.1170-1250), better known as Fibonacci. This series is an addictive progression, beginning with 0,1,1,2,3,5,8,13,21,34,55, etc, where each term after the first two is the sum of the two immediately preceding terms, has numerous fascinating and beautiful properties, and not only governs the ratios of many dimensions in Megalithic geometry, but also those in many natural phenomena: e.g. the laws governing the multi-reflection of light through mirrors, the rhythmic laws of gains and losses in the radiation of energy, the ratio of males to females in honey bee hives and the breeding pattern of rabbits – a symbol of fecundity.

They must also have discovered the use of square roots, deriving from ratios used in planning many of the layouts; e.g. the ratio of the base of an equilateral triangle to its height (1: sq.rt.3); the ratio of the side to the diagonal in a 1:2 rectangle (1: sq.rt.5), etc. These square roots are a type of non-recurring irrational decimal which can never be reduced to an exact quantity. In the Pythagorean number cosmology of ancient Greece, these incommensurable ratios symbolise the immeasurable archetypes that are *a priori* to the visible world of form. The Pythagorean school visualised an organisational foundation for the universe, emerging from invisible vibrational patterns and numerical proportions into the visible world of shape and form. So Megalithic proto-Britons and Bretons were employing Pythagorean mathematics 2,000 years before the birth of Pythagoras.

Yet again, as Anne Macaulay proved, they must have discovered the Golden Mean, the most mysterious and useful of these incommensurable proportional relationships, springing from the pentagon. It is represented by 1.618034, symbolized by *phi* – and most easily recognized as the ratio of each of the longer sides to the shortest side in a triangle in which the angle at the narrow end is 36° and each of the other two angles is 72°. In this unique instance, the smaller part relates to the larger precisely as the larger relates to the whole. *Phi* governs universal laws of proportion: discernible in marine shells, spiralling plants (phyllotaxis), the fuselage/ wingspan ratio of a jumbo-jet, the features of classical architecture and the proportions of the human frame – as famously illustrated in Da Vinci's drawing of Vitruvian Man. It is normally specified as 'y'/2 where 'y' = the sq.rt. of 5 plus 1. Thus, (2.23607 + 1) / 2 = 1.618034. However, *phi* is often invoked by its inverse: (2.23607 – 1) / 2 = 0.618034. Note that **1.618034 x 0.618034 = 1.00** and **1.618034 x 1.618034 = 2.618034.**

In any Fibonacci series, therefore, each new number – the sum of the previous two – divided by the last one, produces a number close to *phi*, and furthermore the greater the numbers the closer the product grows towards *phi*. So, 5/3 = 1.6666, 21/13 = 1.615, 144/89 = 1.618, 233/144 = 1.6180555, etc. (The same symmetrical and harmonic patterns have been recognized in modern times in atomic structures, quantum and wave mechanics, hydrodynamics, electrochemical reactions and molecular bonding.)

Anne Macaulay realized – by one of her many inspirations – that:
3 MR (20.4 ft.) x *phi* (1.618034) = 33 ft. = 2 imperial rods = 11 yards. This half-chain, therefore, stretches back some 6,000 years!

So Anne Macaulay established that our Megalithic ancestors were competent in the Pythagorean *quadrivium* of arithmetic (number), geometry (number in space), music (number in time) and astronomy / astrology (number in space and time). Imperial measures grew from the roots of Europe's earliest civilization. To quote from Boyer & Merzbach's ' A History of Mathematics' (John Wiley & Sons, 1991): "That the beginnings of mathematics are older than the oldest civilization is clear." And from B L Van der Waerden's classic 'Geometry and algebra in ancient civilizations': "We have seen so many similarities between the mathematical and religious ideas current in neolithic England, in Greece, India and China (in the Han period), that we are bound to postulate the existence of a common metrological doctrine from which all these ideas derived." It is this doctrine, the very heart of human culture, that those intent on imposing a metric monopoly are determined to destroy.

Cosmic numerology and customary measure

Multiplying the length of the average nautical mile of 6,076.84 ft by the number of minutes in a circle, 21,600, produces a value for the length of the meridian, the great circle through the poles, of 131,259,744 ft compared to the officially recognized length of 131,332,607 ft . So there are variations, because 'averages' are dangerous and the Earth is an irregular shape. John Michell, however, in his great book, *The Dimensions of Paradise* (Thames & Hudson, 1988) seeks to demonstrate that all cosmological distances are essentially duodecimal multiples which are common to all ancient sciences. He takes 6,082.56 ft, the correct figure for the 'average' mile, times 21,600 = 131,383,296 ft, which equals: 12x12x12x12x12x12 x 44 ft
135,000,000 Roman feet of 0.9732096 ft
90,000,000 Roman cubits of 1.4598144 ft
216,000 Roman furlongs of 608.256 ft
27,000 Roman miles of 4,866.048 ft
129,600,000 Greek ft of 1.01376 ft
86,400,000 Greek cubits of 1.52064 ft
207,360 Greek furlongs of 633.6 ft
25,920 Greek miles of 5,068.8 ft
114,048,000 Egyptian ft of 1.152 ft
76,032,000 Egyptian cubits of 1.728 ft

He points out, moreover, that 131,383,296 ft = 24,883.2 miles which equals (12 to the power 5) / 10 and is decidedly closer to modern estimates from satellite data of the Earth's meridian than the estimate made of that distance by the French scientists of the 18th century for the purpose of defining the metre! The metrologist A R Berriman wrote that, had the French accepted Cassini's proposal in 1720 for a scientific foot based on 1/6,000th part of a minute of average latitude, that unit would have been recognized as identical with the longer Greek foot by which the Parthenon was built. [Though, of course, the Parthenon wasn't built at 'average' latitude; and it could be argued diabolically that the 'English' mile gained some 280ft by Elizabeth I's Act of 1593 – but see APPENDIX VII] However, the fundamental point made by Berriman was that "The names, Greek, Roman and Egyptian are applied to these units by convention only, for they all belong to the same number system and represent fractions of the Earth's dimensions."

Nevertheless, it must be conceded that these findings are open to criticisms that scant scientific evidence exists for the actual use of some of these units within their respective ancient civilizations, that it is impossible to measure any of these units to one millionth part of a foot and that this whole area of study lends itself to historical fantasy or even obsession. But, as explained in one of the articles under 'Standard linear measures', this is not an exercise in physical investigation but the working out of a cosmic scheme. Besides, (a) if calculations are to be illustrated, they have to be expressed precisely, even to an extent that in practice is unrealistic, in order to demonstrate the perfect accuracy of the theory, and (b) ignorant criticism of isolated facets of this vast subject proves little, and (c) yes, the numerology of metrology does become obsessive, but so does dedication to any art or religion, and what subject is more important than the very source of civilization?

The point is that all the ever-increasing evidence that does exist supports John Michell's general conclusions about the inter-relationships of ancient units of measurement, their common derivation from a duodecimal system unified by the English foot, and their consistency with the cosmic dimensions of distance and time.

To quote Michell further: "As calculated above, the longer Roman foot was equal to 0.9732096 ft and the longer Greek foot to 1.01376 ft. Five thousand of these units made up their respective miles, so the Roman mile was of 4,866.048 ft and the Greek mile of 5,068.8 ft. The ratio between them is 24:25, and this ratio is also found to obtain between the Greek mile and the English mile of 5,280 ft. The three mile units thus form a geometric progression.

Roman mile x 25/24 = Greek mile   Greek mile x 25/24 = English mile

It is so unlikely that this neat progression could have arisen by chance that the author feels justified in claiming it as confirming the unit values previously arrived at by other considerations. By linking the English units with those of classical metrology it also justifies making the English foot the prime unit of reference. Modern researchers who accept the fiction that the English units are of recent origin, and adopt the habit of expressing the values of the old measures in terms of the new-fangled metre, thereby disguise from themselves the most significant aspect of ancient metrology, its basis in canonical or classic number – i.e. number that embodies knowledge.

Were it not for that aspect, the business of establishing the exact values of the old units would be of merely academic interest. The fact that those units, as here calculated in English feet, exhibit the same scale of numbers as found in ancient music and geometry is what makes the system of ancient metrology so relevant to the study of traditional science.

As seen above, the English units relate numerically to the Earth's dimensions through the powers of 12: but the most direct geodetic reference to the foot is the equator. In ancient China and Babylon the circle of the equator was divided into 365¼ degrees to represent the number of days in a solar (tropical) year – now taken as 365.2422 days. Each of the 365¼ degrees would measure some 360,000 ft, each minute 6,000 ft and each second 100 ft. If the number of the days in the year is taken as 365.24322, these figures become exact and accord with the estimate of the equator, 131,484,632 ft. Alternatively, dividing the equator into 360 degrees makes each degree equal to 365,243.22 ft, or the number of days in a thousand years.

Another relic of the archaic tradition that produced these divisions of time is our present system of measurement by units of feet, furlongs and miles, with the acre as the unit of land measuring. Those measures, which are still found the most convenient today, were canonized and held sacred, because not only do they relate both to the human and to the astronomical scale, expressing the unity between the macrocosm and the microcosm, but they bring out the same numbers in the dimensions of the solar system as were given to the units of time. The canonical dimensions of the earth, sun and moon, are:

Diameter of the sun	864,000 miles	(12 x 12 x 6,000)
Radius of the sun	432,000 miles	(12 x 12 x 3,000)
Diameter of the moon	2,160 miles	(6 x 6 x 60)
Radius of the moon	1,080 miles	(6 x 6 x 30)
Mean diameter of earth	7,920 miles	(12 x 660)
Mean circumference of the earth	24,883.2 miles	(12 x 12 x 12 x 12 x 1.2)
Distance from earth to moon	237,600 miles	(6 x 60 x 660) or 60 x earth's radius
Distance from earth to sun	93,312,000 miles	(6 to the power 6 x 2,000)"

The sheer force of Michell's general theory – the framework surrounding these celestial distances whose measurements in miles factorize by 6 and 12 – is compelling. Consider, also, these additional facts that he deduced:

    3,168 ft = 1,164 megalithic yards of 2.722 ft
    31,680 ft = 6 miles
    31,680 inches = ½ a mile
    31,680 furlongs = 3,960 miles = radius of the earth [and so on]

(adapted from an article contributed to *The Yardstick*, the BWMA journal, by Robin Heath, author of four books, including '*Sun, Moon & Stonehenge*', and '*A Beginner's Guide to Stone Circles*' (Hodder, 1999)

"If one assumes that the megalithic yard represents a lunation period – the time between two new moons – then the length of 1.ft (12") marks the required calendrical period between the end of the lunar year (which comprises 12 lunations taking 354.367 days) and the end of the solar year which comprises 365.242 days. Assuming that the MY was the primary unit, then the derivative foot appears to have formed a logical and essential part of the astronomical researches of our neolithic ancestors. If, however, the foot *preceded* the MY (perhaps through the fact that 1,000th of 1 degree of arc around the equatorial circumference equals 365.242 ft), then knowledge of the roundness of the earth must predate use of the MY – earlier than 3,000BC. It would be logical for the key calendrical time interval – the 10.875 days elapsing between the ends of the lunar and solar years – to have been represented by an existing unit of length. It does appear that the foot of 12" was adopted accordingly.

My story doesn't end here, for there are 12.368 lunations in a year. Again assuming that 1 MY represents a lunation, then to tot up the exact number of lunations in the year, all one has to do is add one inch to a MY each and every time one observes a lunation. After the lunar year, the inches add up to 0.368 lunations, exactly the required over-run. When demonstrating accurate calendrical predictions with my student groups, I use a plastic Woolworth's foot-ruler, marked in inches, to predict lunations and eclipses precisely to the day, years in advance. Not very megalithic, perhaps, but highly accurate and apparently using those same measures of antiquity now rendered illegal by the government of the same lands that originally built those great stone circles.

Many of the greatest metrologists have suggested that the primary units of length were derived from units of time. They all concluded that such measures were so strictly defined and so rigorously organized that they must have a basis on some absolute natural measure. I submit that this 'natural measure' was the lunation period. But whatever *length* a system of weights and measures was based upon, there was a coordinating link to units of *time*. The road to effective weights and measures begins with time measurement, leads to length, and finally to areas, volumes and weights. Accounts of the MY totally confirm this thinking; the cycles of the sun and moon defining the *absolute natural measure*.

So the MY may be considered a calendrical analogue of the lunation period, the foot (and the cubit*) being proportioned within it to reveal the length of the solar year, following the twelfth lunation. The accuracy is astonishing, to within less than half an hour in one solar year. The conclusions from this research have enormous implications for archaeology and the understanding of human prehistory. They confirm the primary connection between mensuration and the moon. They also clearly offer one plausible explanation for the historical basis of the foot and its twelve-fold division into inches."

---

*Robin Heath provides further corroboration by showing that 1 Megalithic Yard (2.72 ft) *exactly equals* 1 ft *plus* 1 ancient Egyptian 'Royal Cubit' of 1.72 ft, but we cannot expand upon that here.

EPILOGUE

The metric system as an alternative to, or to replace, the imperial system?

1   A metric system, in which all measurements should be multiplied or divided by ten, was first devised by a priest of Lyons, Gabriel Mouton (which aptly means 'sheep') in 1670, although his idea was to take a one-thousandth part of the nautical mile (about 6 ft) as the unit of length. The *metre* was introduced by French revolutionaries in 1795, calculated (inaccurately) as a ten-millionth part of the distance of a line running on the Earth's surface through Paris, from the North Pole to the Equator. The relevance of that line to human measurement or practical purposes remains unclear; but it was a consideration that the French were envious of British maritime supremacy and wanted the prime meridian of longitude to run through their capital rather than through Greenwich. This far-fetched formula produced an arbitrary unit which was, very neatly, just larger than the detested English yard.

2   A standard metre bar, made of platinum-iridium alloy, was kept in Paris. The length was defined in 1960 in terms of optical wavelengths in a vacuum of radiation from atoms of krypton-86; but even that proved too simple and unreliable, so it was next expressed by reference to the wavelength of the radiation from an iodine-stabilized helium-neon laser, before finally resorting to measurement in terms of the speed of light; all of which is very handy to know.

3   From the metre as a unit of length, the French developed the *are* as the unit of area (i.e. 100 square metres or 1 square decametre), the *stere* as the unit of cubic measure (i.e. 1 cubic metre), the litre as the unit of capacity (1/1,000th of a *stere*) and the gram as the unit of mass of one cubic centimetre (1/1,000th of a litre) of water. As the French tended to use centimetres rather than metres for everyday purposes, and also retained the second as the standard measure of time (even though the clock and calendar are wholly non-metric), this spectrum of units became known as the centimetre-gram-second or CGS system: which was actually the brainchild of Lord Kelvin. But by the beginning of the 20th century the metre and kilogram (the weight of one litre of water) had become the international metric units of length and mass respectively, and so the MKS system was adopted by the Conference Generale des Poids et Mesures. In 1950 the ampere (measuring electric current) was recognized as a fourth basic unit, then in 1960 at the 11th CGPM the kelvin (temperature) was included (273°K = 0° Celsius = 32°F) as well as the candela (luminous intensity), radian and steradian

to constitute the Système International d'Unités (SI), which was further extended in 1964 at the 12th CGPM to include radionuclides, and yet again at the 14th CGPM in 1971 when the seventh base unit, the *mole*, was adopted.

4   How strange that Britain and America have managed to muddle along, ever since standardization of the Imperial system in 1825, without holding international conferences at all! The result of these frequent metric conferences, in contrast, so far from producing one universal system, has been to leave many metric countries behind at different stages in this process of forced evolution, and for most countries to develop hybrid units which amalgamate metric and their own indigenous measures, adding further to the complexity and confusion of the global system. In Britain we already have the 'metric foot' of 30cm, equal to 11.81 inches [see ***footnote B*** above and para. 12 below]

In contrast, BWMA rulers show, on one edge, inches divided into 1/3, 1/6, 1/12 & 1/24, another edge ½, ¼, 1/8, 1/16 &1/32, and on the other side 1/10 & 1/20 on one edge – for these are just additional fractions – and on the fourth edge the popular mapping scale of 1¼ inch to the mile, which is very close to 1:50,000 natural. Inherent complexity and confusion throughout the metric system arises from the distinction between the use of Latin prefixes for sub-units (decimetre, centimetre, etc) and Greek prefixes for multiple units (decametre, kilometre, etc). Indeed, there are 16 prefixes alone (8 Latin and 8 Greek) and symbols for each one: some in lower case and others in upper – so that if a lower case 'm' is used instead of 'M', the error could be by a factor of a thousand million.

5   One of the few American champions of metrication asked in the journal *Mathematics Teacher:* "If the average horse is 15.84 hands tall, and we know that one mile measures 15,840 hands, is one mile equal to 1,000 horses? And that's just for the statute mile: how about sea horses and the nautical mile?" But the respect earned by his humour was lost when he complained that "Fractions are difficult for the French. The only fraction the French use is one-half. They will order a demi-litre of wine, but a smaller amount is not a quarter litre (2.5 decilitres) but rather an even 2.0 decilitres.. [Not so: *un quart de vin* (25cl) is common, and so are quarter bottles of 18.75cl] Our (US and UK) fractional system is the awkward octal or hexadecimal one of computers; whereas theirs goes 2, 1, 0.5, 0.2, 0.1, 0.05, and so on, to form a sequence that repeats in 10s." So it is the arithmetic of computers that is at fault, not the metric system; and ignorance of fractions is a sign of superior intelligence!

6   To quote Derek Turner (Editor of *Right Now!*): "Technocrats are not susceptible to the visceral urges felt by us lesser mortals. Not for them irrational appeals to tradition, nation, family, love, honour, property, competition, music or metaphysics. They are very much like their spiritual predecessors in Revolutionary France, who devised the metric system in the interests of greater efficiency and as a means of divorcing the French from their glorious past. There are discernible connections between the clicking of slide-rules and the rattling of tumbrils. The American writer Bill Kauffman, in one of his superb essays, shows that he understands the links between the phenomena of internationalism and metrification. Among the unsung patriots of our day, my everyday heroes, are the ornery old men who speak of quarts, not litres, the refractory kids who flunk tests on their metric conversion tables, and the track officials who still stage 100-yard dashes and mile runs. There is nothing inevitable about global markets, global culture or global government, in a world where the number of independent nations has increased five-fold since 1950."

7   Furthermore, the United Nations publishes a 138-page handbook, '*World Weights and Measures*', detailing national currency and measurement systems. This volume destroys the myth of global metrication. It reveals not only the surprising number of countries still using the pint-foot-pound system but also how widespread, within countries that have long been officially metric, is the continuing use, for everyday purposes, of their indigenous units. The book recognizes that: "It would not be practical to show, for each of the thousands of units currently in use, equivalents in all the other systems of measurement." Indeed, it lists about 1,675 non-metric units, their values and countries of origin. So, ironically, it is the monolithic UN that provides the best possible proof that the metric system has a far from firm grip on the world.

8   Surely the most stupid and perverse example of metrication was by Radio 4, announcing the news of the superb achievement by the British *Thrust SSC* team in raising the world land-speed record past Mach 1 – on the sands of the Nevada Desert – to "1,149 kilometres per hour". Never mind that in Britain it remains legal to use 'mph' as we continue to do universally; never mind that in the USA 'kph' is meaningless: here we have the BBC, which proclaims its commitment to education, referring to a speed in 'kph' over the prescribed distance of the ***measured mile***!

9   To quote John Strange of BWMA: "The measurement of angles, by the arcs they cut on a circle, is as old as the notion of angle itself, and was already

known to the Babylonians [and Megalithic Britons] 4,000 years ago. They divided the circle into 360° and introduced the sexagesimal scale. 60 was probably chosen because it is divisible by 2, 3, 4, 5 and 6 (as well as 10, 12, 15, 20 and 30). So when the Greeks required a finer division of the circle, they naturally divided each degree into 60 minutes. We still use these units today, despite French efforts to decimalize angles. (The French tried dividing the right angle into 100 grades or *gons* but the system failed because the sexagesimal system was so well rooted, both for geometry and for measuring time).

The metric unit for measuring angles is the radian. Two right angles equal 200 *gons*, 180° or 3.14159265 radians *(pi)*. There is, however, a technical difficulty. As the Greeks recognized, angles are not true magnitudes and the methods developed in Book V of Euclid's *Elements* cannot be applied to them. Suppose for a moment that the Earth were a perfect sphere. The meridians are semi-circles, drawn on its surface, whose end points are the North and South Poles; the polar axis is the straight line which runs through the centre of the Earth, joining both poles. Then the nautical mile is the distance between two points on the same meridian whose latitudes differ by one minute; so the distance along a meridian from either of the poles to the equator must be 90 times 60 or 5,400 nautical miles.

However, because the Earth is squashed at the poles, its curvature is greater near the equator than elsewhere and consequently the nautical mile, as defined above, varies from nearly 6,046 ft near the equator to about 6,107 near the poles. The Admiralty decided that it should be fixed at 6,080 ft but the international nautical mile has since been defined as 1.852 km which is just under 6,076' 1½". The French wanted to define the metre in terms of the distance from the North Pole to the equator measured along the Paris meridian. This distance was to be 10 megametres so that any two points on the meridian, whose latitude differed by one grade, would be 100 km apart. Each grade is divided into 100 parts and the kilometre was to replace the nautical mile. But the Earth is divided by meridians into 24 time zones. The two meridians which bound one of these time zones meet at the poles at an angle of 15° (360°/ 24). This relationship between angles and time zones is rather awkward if the angles are measured in grades. The French had foreseen the difficulty and, on 5 October 1793, the Convention decreed the decimalization of time. This proved a failure and a further decree on 7 April 1795, the date of birth of the metric system, suspended the operation of the earlier one indefinitely. So, in a temporal sense, the metric system was still-born. **As a result, traditional units for time are still used universally and navigators still use nautical miles.**

Scientists, who do not mind if the unit they are using is sometimes disproportionate to the thing they are measuring, are quite happy to use the *second* for time and the *metre* for distance."

10 From '*Men and Measures*' by Edward Nicholson (1912): "In England a few genuine enthusiasts, and many more who have caught the scientific and cosmopolitan craze, take to the metric system as they take to learning Esperanto, and so long as they have not to use the one for business or the other in conversation, their enthusiasm lasts, especially when it affords opportunities for showing themselves friends of science and progress. But when the contagion spreads so wide that it threatens to revolutionize the habits and customs of a nation and its whole manufactures and trade, the danger is most serious. *The favour which the metric system has found amongst a small proportion of English people is largely due to their ignorance of their own system, an ignorance very excusable when there exists no official statement of our system, or even of its standards.*"

11 From a Conference Paper delivered by Professor C Piazzi Smyth (former Astronomer Royal for Scotland) in 1872: "We, as a nation, have done as yet little beyond merely offering a dull resistance to any kind of change in things which we undervalued, but to which we have clung with the inertia of custom and wont, because they had come into our possession we knew not how or when."

12 The British Standards Institute has adopted 30 cm as a building module, following the disuse of the decimetre; but now even the cm is slowly dying, to leave only the metre and millimetre. By mere coincidence (as Arthur Whillock has pointed out), 30cm is very close to the original Greek Attic foot of sixteen digits; enlarged from the Pythic foot of 12 digits. [see para. 4 above]

Whillock commented (1973): "This new metric foot, with its twelve 'inches' of 25mm (used for timber sizes, bolt lengths, tape widths, etc), has caused alarm in some quarters, but give it time; the metre may become the new 'ell' of 40 inches! Relieved of the need to support technical work for the time being, our foot will be free to adjust itself to the value used by the designer of the Parthenon, and it is fitting that his descendants are engaged in putting measurement back onto the rails it left some 150 years ago."

13 To quote again from Jacob Bronowski's '*The Ascent of Man*': "Take a beautiful cube of pyrites. Or to me the most exquisite crystal of all,

fluorite, an octahedron. (It is also the natural shape of the diamond crystal.) Their symmetries are imposed on them by the nature of the space we live in – the three dimensions, the flatness in which we live. And no assembly of atoms can break that crucial law of nature. Like the units that compose a pattern, the atoms in a crystal are stacked in all directions. So a crystal, like a pattern, must have a shape that could extend or repeat itself indefinitely. That is why the faces of a crystal can only have certain shapes; they could not have anything but the symmetries in the pattern. For example, the only rotations that are possible go twice or four times for a full turn, or three times or six times – not more. And not five times. You cannot make an assembly of atoms to make triangles which fit into space regularly five at a time."

14 To quote from an article in *Nature* published in 1922: "In view of the vigorous and sustained efforts of the exponents of the metric system, and the eminent names that are to be found among them, it is perhaps not a little surprising that it makes so little progress towards general acceptance in Great Britain. **The Weights and Measures Act of 1897 legalized the metric denominations for use in trade, and was expected to lead to its advantage being so generally recognized that the Imperial system would soon disappear. Twenty-five years have now elapsed and the position is almost unchanged. In fact, the policy of compulsory introduction of the metric system by law, which formerly was always strongly supported, was ruled out by the Metric Committee of the Conjoint Board of Scientific Societies in its Report of 1919**....The subject of compulsion is not likely again to be seriously considered for some time at least."

15 To quote again from John Neal's *'Opus 2 – All done with Mirrors':* "The universal acceptance of the metric system was assured by virtue of its convenience to the international banking regime, all other considerations being subservient to its requirements. The primary reason that the United States has resisted conversion for so long is that its monetary system was decimalised from the very foundations of the Nation. With its international currency trading already conformed, no necessity was seen to interfere with its internal standards. Modern money itself being artificial, it is perfectly suited to an artificial counting method that can only express quantity; whereas in the fields of the arts and sciences, a system of harmonic proportions among quantitative expressions is ideally satisfied by the traditional measures.

The metre, when legalised by the British in 1864 for use in contracts, was resolved to be 39.3708 inches. Yet when legalized in 1897 for purposes of general trade, it was more accurately standardized as 39.370113 inches. But when formally defined for recognition by the United States in 1866, there it was 39.37 inches. Again, when the metre was adopted in Japan in terms of the *shaku*, the definition expressed is also minutely at variance with both the British and US standards. Each nation has a slightly different interpretation. This lack of agreement on conversion rates can have disastrous consequences for scientific endeavours involving time and distance.

APPENDIX I

A few foreign units of linear measure corresponding to the Imperial foot. In every case, the divisions of the unit are twelfths, like the inch.

COUNTRY	NAME	LENGTH (inches)
Turkey	½ pik	13.5
Portugal	pe	12.96
France	pied	12.79
Austria	fuss	12.44
Germany	fuss	12.36
Norway & Denmark	fod	12.36
Babylonia (ancient)	(?)	12.24
Roman	pes	11.60
Greece	pous	12.08
Russia	foute	12.00
Japan	shaku	11.93
Belgium	pied	11.81
Holland	voet	11.15

Continental Mile Measurements

Austrian mile	8,296 yards
Spanish mile	5,028 yards
Russian verst	1,167 yards
Italian mile	1,467 yards
Polish mile	4,400 yards

APPENDIX II

Some quotations from Shakespeare

Contributed by John Constable

Faith, here's an equivocator, that could swear in both the scales against either scale   (*Macbeth: Act II Scene iii line 9*)

Now would I give a thousand furlongs of sea for an acre of barren ground (*The Tempest: I. i. 70*)

With one soft kiss a thousand furlongs ere / With spur we heat an acre (*Winter's Tale: I. ii. 94*)

My bosky acres, and my unshrubbed down   (*The Tempest: IV. i. 81*)

Between the acres of the rye, / With a hey, and a ho, and hey-nonny-no (*As You Like It: V. iii. 24*)

Over whose acres walked those blessed feet
Which fourteen hundred years ago were nailed,
For our advantage, on the bitter cross   (*1 Henry IV: I. i. 25*)

And if thou prate of mountains, let them throw
Millions of acres on us   (*Hamlet: V. i. 302*)

O coz, coz, coz, my pretty little coz, that thou didst know how many fathom deep I am in love!   (*As You Like It: IV. i. 218*)

Full fathom five thy father lies   (*The Tempest: I. ii. 394*)

Where fathom-line could never touch the ground   (*1 Henry IV: I. iii. 204*)

And not the worst of the three but jumps twelve foot and a half by the squier [square]   (*The Winter's Tale: IV. iii. 349*)

Three foot of it doth hold: bad world the while!   (*King John: IV. ii. 99*)
If I travel but four foot by the square further afoot, I shall break my wind
(*1 Henry IV: II. ii. 13*)

By the good gods, I'd with thee every foot   (*Coriolanus: IV. i. 56*)

Now would I give a thousand furlongs of sea for an acre of barren ground
(*The Tempest: I. i. 70*)

Item: sack, two gallons   (*1 Henry IV: II. iv. 595*)

That you should have an inch of any ground
To build a grief on   (*2 Henry IV: IV. i. 109*)

I'll show thee every fertile inch o'the island   (*The Tempest: II. ii. 160*)

I will fetch you a tooth-picker now from the furthest inch of Asia
(*Much Ado: II. i. 277*)

I'll not budge an inch, boy   (*Taming of the Shrew: Induction. i. 14*)

One inch of delay more is a South Sea of discovery
(*As You Like It: III. ii. 207*)

For every inch of woman in the world,
Ay, every dram of woman's flesh is false   (*The Winter's Tale: II. i. 136*)

I have speeded hither with the very extremist inch of possibility
(*2 Henry IV: IV. iii. 38*)

Being now awake, I'll queen it no inch further
(*The Winter's Tale: IV. iii. 462*)

My inch of taper will be burnt and done   (*Richard II: I. iii. 223*)

Not an inch further   (*1 Henry IV: II. iii. 119*)

Ay, every inch a king   (*King Lear: IV. vi. 110*)

Am I not an inch of fortune better than she?
(*Antony and Cleopatra: I. ii. 61*)
For I'll cut my green coat, a foot above my knee,
And I'll clip my yellow locks, an inch below mine eye
(*The Two Noble Kinsmen*)   [*authorship disputed*]

Whom I, with this obedient steel – three inches of it –
Can lay to bed for ever   (*The Tempest: II. i. 291*)

Ask them how many inches / Is in one mile: if they have measur'd many,
The measure then of one is easily told.   (*Love's Labour's Lost: V. ii. 189*)

Am I but three inches? Why, thy horn is a foot; and so long am I at the least.
(*Taming of the Shrew: IV. i. 29*)

Eight yards of uneven ground is threescore and ten miles afoot with me
(*1 Henry IV: II. ii. 27*)

And buckle in a waist most fathomless
With spans and inches so diminutive   (*Troilus and Cressida: II. ii.30*)

As many inches as you have oceans   (*Cymbeline: I. ii. 21*)

My sweet ounce of man's flesh!   (*Love's Labour's Lost: III. i. 142*)

Nay then, I must have an ounce or two of this malapert blood from you
(*Twelfth Night: IV. i. 48*)

Give me an ounce of civet, good apothecary, to sweeten my imagination
(*King Lear: IV. vi. 132*)

Weigh you the worth and honour of a king / So great as our dread father in a scale / Of common ounces?   (*Troilus and Cressida: II. ii. 26*)

Good faith, a little one; not past a pint, as I am a soldier   (*Othello: II. iii. 69*)

The pound of flesh which I demand of him   (*Merchant of Venice: IV. i. 99*)

Three pound of sugar; five pound of currants…four pound of prunes
(*The Winter's Tale: IV. ii. 40*)

Butter at eleven pence a pound   (*Sir Thomas More*) *[authorship disputed]*
Indeed, I am in the waist two yards about (*Merry Wives of Windsor: I. iii. 43*)

I looked a' should have sent me two-and-twenty yards of satin
(*2 Henry IV: I. ii. 48*)

APPENDIX III

From the 'REPORT from the SELECT COMMITTEE on WEIGHTS AND MEASURES (communicated by the Commons to the Lords)' 1816:

"Your Committee, in the first place, proceeded to enquire what measures had been taken to establish uniform Weights and Measures throughout the Kingdom. They found that this subject had engaged the attention of Parliament at a very early period. The Statute Book, from the time of Henry the Third, abounds with Acts of Parliament enacting and declaring that there should be one uniform Weight and Measure throughout the Realm; and every Act complains that the preceding Statutes had been ineffectual, and that the Laws were disobeyed.

In order to obtain some information as to what were the best means of comparing the standards of length, with some invariable natural standard, your committee proceeded to examine…..From the evidence of these gentlemen, it appears that the length of a pendulum making a certain number of vibrations in a given portion of time, will always be the same in the same latitude; and that the standard English yard has been accurately compared with the length of the pendulum, which vibrates 60 times in a minute in the latitude of London. The length of this pendulum is 39.13047 inches, of which the yard contains 36. Any expert watch-maker can easily adjust a pendulum, which shall vibrate exactly 60 times a minute.

The French Government have adopted as the standard of their measures, a portion of an arc of the meridian, which was accurately measured. The standard metre, which is the 10,000,000[th] part of the quadrant of the meridian, which is engraved on the Platina Scale preserved in the National Institute, has been compared with the English standard yard .and was found to exceed it, at the temperature of 32 degrees, by 3.3702 inches; and at the temperature of 55 degrees, by 3.3828 inches. The standard yard may therefore be at any time ascertained, by a comparison either with an arc of the meridian, or the length of a pendulum, both of which may be considered as invariable.

The standard of linear Measure being thus established and ascertained, the measures of capacity are easily deduced from it, by determining the number of cubical inches which they should contain. The standard of weight must be derived from the measures of capacity, by ascertaining the weight of a given bulk of some substance of which the specific gravity is invariable. Fortunately that substance which is most generally diffused over the world, answers this condition. The specific gravity of pure water has been found to be invariable at the same temperature; and by a very remarkable coincidence, a cubic foot of pure water (or 1,728 cubical inches) at the temperature of 56½ degrees by Fahrenheit s thermometer, has been ascertained to weigh exactly 1,000 ounces Avoirdupois, and therefore the weight of 27.648 inches is equal to one pound Avoirdupois.

This circumstance forms the groundwork of all succeeding observations of Your Committee. Although in theory the standard of weight is derived from the measures of capacity, yet in practice it will be found more convenient to reverse this order. The weight of water contained in any vessel, affords the best measure of its capacity, and is more easily ascertained than the number of cubical inches by gauging. Your Committee therefore recommends that the measures of capacity should be ascertained by the weight of pure or distilled water contained by them, rather than by the number of cubical inches, as recommended in the 4th Resolution of the Committee of 1758.

Your Committee is also of opinion, that the standard gallon, from which all the other measures of capacity should be derived, should be made of such a size as to contain such a weight of pure water of the temperature of 56½ degrees, as should be expressed in a whole number of pounds Avoirdupois, and such also as would admit of the quart and pint containing integer numbers of ounces, without any fractional parts. If the gallon is made to contain 10 pounds of water, the quart will contain 40 ounces and the pint 20. If this gallon is adopted, the bushel will contain 80 lbs of water, or 2,211.84 cubical inches; the quart 69.12 cubical inches….the pint 34.56 cubical inches (which is exactly $1/100^{th}$ part of a cubical foot) …."

From the 'FIRST REPORT of the COMMISSIONERS appointed to consider the subject of Weights and Measures' 1819:

"III The Subdivisions of Weights and Measures, at present employed in this Country, appear to be far more convenient for practical purposes than the Decimal Scale, which might perhaps be preferred by some persons, for making calculations with quantities already determined. But the power of expressing a third, a fourth, and a sixth of a foot in inches, without a fraction, is a peculiar advantage of the Duodecimal Scale; and for the operations of weighing and of measuring capacities, the continual division by Two renders it practicable to make up any given quantity, with the smallest possible number of standard Weights or Measures, and is far preferable, in this respect, to any decimal scale. We would therefore recommend, that all the multiples and subdivisions of the Standard to be adopted should retain the same relative proportions to each other, as are at present in general use."

From the 'THIRD REPORT of the COMMISSIONERS' 1821:

"…And we have found by the computations…that the weight of a cubic inch of distilled water, at 62 degrees Fahrenheit, is 252.72 grains of the Parliamentary Standard Pound of 1758, supposing it to be weighed in a vacuum. We beg leave therefore finally to recommend, with all humility, to Your MAJESTY, the adoption of the Regulations and Modifications suggested in our former Reports; which are principally these:

That the Parliamentary Standard Yard, made by Bird in 1760, be henceforwards considered as the authentic legal Standard of the British Empire; and that it be identified by declaring, that 39.1393 inches of this standard [of which the yard contains 36] at the temperature of 62 degrees Fahrenheit, have been found equal to the length of a Pendulum supposed to vibrate seconds in London, on the level of the sea, and in a vacuum.

That the Parliamentary Standard Troy Pound, according to the two pound weight made in 1758, remain unaltered; and that 7,000 Troy Grains be declared to constitute an Avoirdupois Pound; the cubic inch of distilled water being found to weigh at 62 degrees, in a vacuum, 252.72 Parliamentary grains.

That the Ale and Corn Gallon be restored to their original equality, by taking, for the statutable common Gallon of the British Empire, a mean value, such that a gallon of common water may weigh 10 pounds avoirdupois in ordinary circumstances, its content being nearly 277.3 cubic inches; and that correct Standards of this IMPERIAL GALLON, and of the Bushel, Peck, Quart and Pint, derived from it, and of their parts, be procured without delay for the Exchequer, and for such other offices in Your Majesty's dominions, as may be judged most convenient for the ready use of Your Majesty's subjects."

APPENDIX IV

From the Report submitted to the US Congress in 1821 by **John Quincy Adams**, *then Secretary of State, prior to his election as President in 1825. He had been commissioned to produce this Report in order to help Congress determine the system of weights and measures they should adopt for the new Nation. It is a most impressive document, reflecting those same fundamental truths concerning the origins and current condition of the imperial system, and also the deficiencies of the metric, that still apply today.*

"Considered as a whole, the established weights and measures of England are but the ruins of a system, the decays of which have been often repaired with materials adapted neither to the proportion nor to the principles of the original construction. The metrology of France is a new and complicated machine, formed upon principles of mathematical precision, the adaptation of which to the uses for which it was devised is yet problematical, and abiding, with questionable success, the test of experiment.

To the English system, belong two different units of weight and two corresponding measures of capacity, the natural standard of which is the difference between the specific gravities of *wheat* and *wine*. To the French system, there is only one unit of weight and one measure of capacity, the natural standard of which is the specific gravity of water.

The French system has the advantage of unity in the weight and the measure, but has no common test of both: its measure gives the weight of water only. The English system has the inconvenience of two weights and two measures; but each measure is, at the same time, a weight. Thus the gallon of wheat and the gallon of wine, though of different dimensions, balance each other. A gallon of wheat and a gallon of wine each weigh eight pounds *avoir-dupois*. The *litre* in the French system is a measure for all grains and all liquids.; but its capacity gives a weight only for distilled water. As a measure of corn, of wine, or of oil, it gives the space they occupy, but not their *weight*. Now, as the weight of these articles is quite as important in the estimate of their quantities as the space which they fill, a system which has two standard units for measures of capacity, but of which each measure gives the same weight of the respective articles, is quite as uniform as that which, of any given article, requires two instruments to show its quantity – one to measure the space it fills and another for its weight. In the difference between the specific gravities of corn and wine, nature has also dictated two standard measures of capacity, each of them equiponderant to the same weight.

This diversity existing in nature, the Troy and Avoirdupois weights, and the corn and wine measures of the English system are founded upon it. In England it has existed as long as any recorded existence of man upon the island; but the system did not originate there. The weights and measures of Greece and Rome were founded upon it. The Romans had the *mina* and the *libra*, the nummulary pound [relating to coinage] of 12 ounces and the commercial pound of 16. The avoirdupois pound came through the Romans from the Greeks, and through them, in all probability, from Egypt. Of this there is internal evidence in the weights themselves, and in the remarkable coincidence between the cubic foot and the 1,000 ounces avoirdupois, and between the ounce avoirdupois and the Jewish *shekel*; and if the *shekel* of Abraham was the same as that of his descendants, the avoirdupois ounce may, like the cubit, have originated before the flood.

The result of these reflections is that the uniformity of nature for ascertaining the quantities of all substances, both by gravity and by occupied space, is a uniformity of *proportion*, and not of *identity*; that, instead of one weight and one measure, it requires two units of each, *proportioned* to each other; and that the original English system of metrology, possessing two such weights and two such measures, is better adapted to the only uniformity applicable to the subject, recognized by nature, than the new French system which, possessing only one weight and one measure of capacity, identifies weight and measure by only the single article of distilled water; the English uniformity being relative to the *things* weighed and measured, and the French only to the *instruments* used for weight and mensuration."

Later in this Report, Adams commented on the decimal principle. "It can be applied, only with many qualifications, to any general system of metrology; that its natural application is only to *numbers*; and that *time, space, gravity* and *extension* inflexibly reject its sway…It is doubtful whether the advantage to be obtained by any attempt to apply decimal arithmetic to weights and measures, would ever compensate for the increase of diversity which is the unavoidable consequence of change. Decimal arithmetic is a contrivance of man for computing numbers, and not a property of *time, space* or *matter*. Nature has no partialities for the number *ten*; and the attempt to shackle her freedom with it will forever prove abortive."

APPENDIX V

The Gallon   Contributed by John Strange

The gallon illustrates well the difference in outlook between the metric and imperial systems. Referring to Appendix III (particularly the 1821 Report and penultimate para. of the 1816 Report), we see the Commissioners' concern to provide a measure that was not only precise but also practical. So, if you want to find a container's capacity, all you have to do is to weigh it empty and then weigh it again filled with pure water. The difference between the two weights in ounces is the volume of the container in fluid ounces. (For greater precision, the water should be at a temperature of 62°F.)

Let us look first at the French system. How were they to choose a unit of mass? They decided that the unit should be the mass of one cubic decimetre of water. But that mass depends on the temperature of the water – so that had to be decided upon. Now, if the temperature varies slightly from 39.16°F, the density of water hardly changes, so it makes sound theoretical sense to choose that point in order to minimize the effect of any error in temperature, which the

French authorities accordingly did. They made a cylinder of platinum-iridium whose mass was very nearly that of 1 cubic decimetre of water at its most dense...and that is the kilogram. When it was subsequently discovered that the standard kilogram was very slightly heavier than intended, the litre was defined as the volume occupied by 1 kilogram of water at its most dense. This turns out to be very nearly 1,000.028 cubic centimetres – redefined in 1964 simply as 1,000 cu.cm. (The Revolutionaries should not be blamed for miscalculating the metre and the kilogram: As one of them said: "La republique n'a pas besoin de savants", so they guillotined Lavoisier, the discoverer of oxygen.)

It is a matter of common experience that if you venture a short way into the sea and pick up a stone, it feels heavier as soon as it is withdrawn from the water. When it's in the water, it is buoyed up. Archimedes' principle states that: "When a solid body is immersed (wholly or only partly) in a fluid, it experiences an upward thrust equal to the weight of the fluid displaced." Another obvious example is that of a balloon in air. Indeed, even a man experiences an upthrust from the surrounding air, but it's so small – about 3oz – as to be negligible. Now, *nos amis* did not overlook this effect when they weighed their water. So they ended up with a rather impractical definition of the litre. The water had to be about 39.16°F (pretty damn cold) and the weighing had to be performed in a vacuum!

Compare this with the imperial gallon. The temperature is a very reasonable 62°F, and we don't have to worry about the weight of the air displaced because that's taken care of. As the Third Report says – having spoken earlier of the theoretical weighing in a vacuum – "a gallon of common water may weigh ten pounds avoirdupois *in ordinary circumstances*" [my emphasis]. Does that not sum up our customary measures: 'common sense in ordinary circumstances'?

A gallon is today defined as 4.54609 litres – i.e. very nearly 277.4194328 cu.in. or 0.16054365323 cu.ft., while the density of water at 62°F is very nearly 997.68 oz per cu.ft. Consequently, the mass of 1 gallon of water at 62°F is very nearly 997.68 x 0.16054365323 = 160.17119 oz. But that gallon of water experiences an upthrust of 0.19515 oz, this amount being the weight of air it has displaced. So the gallon of water appears to weigh only 159.97604 oz. Yet that's not the end of the story, for the 10 lb in the other scale pan have displaced 0.02393 oz of air. So, finally, the gallon of water appears to weigh 159.99997 oz: you could hardly ask for anything more accurate than that! Therefore, while the French method is theoretically capable of giving greater precision, it is not quite so accurate and less useful in practice, because we are not going to use water at 39.16°F and we are not going to weigh it in a vacuum.

Much of the data for this article is taken from The Weights and Measures Act of 1963

APPENDIX VI

Excerpts from the 'Rules of Lawn Tennis' Contributed by John Strange

"The singles court is 78 ft long and 27 ft wide. The service court is 21 ft long. The doubles court is 36 ft wide. The net posts are 3 ft outside the court. The net is 3'6" high at the posts and 3'0" in the middle. The lines are between 1 and 2 inches wide except the base line which is between 2 and 4 inches wide.

The diameter of the balls is between two-and-a-half and two-and-five-eighths inches and their weight between 2 oz and 2 oz 1 dram. When dropped onto a concrete base from a height of 100 in., the ball shall bounce to a height between 53" and 58". The tests are performed at a temperature of 68°F and an atmospheric pressure of 29.95" of mercury."

Proof Spirit Contributed by John Strange

In the old days, when a ship arrived from France with brandy on board, Excise men put some gunpowder on the quay-side, poured a little of the brandy over it and applied a match. If the fire fizzled out, the brandy was declared to be under proof; if it flared up it was over proof, but if it burned gently it was certified as proof spirit.

Nowadays, a volume of spirits is at proof (i.e. 100°) if it weighs 12/13 of the weight of the same volume of water, both being at a temperature of 51°F. In practice, if we mix 2 pints 9 fl.oz. (3 lb 1oz.) of water with 3 pints (3 lb) of pure alcohol we get about 5¼ pints of proof spirit; for there's some shrinkage when the liquids are mixed.

Spirits sold in the UK are usually 30 percent under proof (i.e. 70 degrees), corresponding to a 40 percent alcoholic content. This 7/4 ratio reflects the fact that pure alcohol equals 175 degrees proof. But the US proof scale differs: a US whisky of 80 degrees proof equating to 70 degrees on the UK scale.

APPENDIX VII

The 'King's Girth' and the cosmological pattern of the Saxon royal court

Contributed by JOHN MICHELL

copyright reserved by the author, who has generously given BWMA permission for first publication here

(John Neal, John Michell's colleague, when sending us this article, commented: *"The greatest metrologist of all, Livio Stecchini, found examples of a wide range of measurements engraved as standards on Roman monuments, in addition to those that we accept as 'Roman'. He stated, 'These monuments confirm that not only in Rome one used the natural version of the Roman foot, but also that in the ancient world all of the units of length formed a system and were used concurrently.' In addition, the practices of the Roman* gromatici, *or planners, who surveyed the towns and consecrated the surrounding area, would appear to be based on the identical canon as the Saxon – therefore, our own. In spite of the fact that the Saxons preserved the Roman estate and land divisions, as in the* ville *and the* tun, *there is no reason to believe that their methods were inherited from a Roman source….the only point on which I would disagree with John Michell's treatment of the evidence is his statement that the necessary knowledge is 'revealed' intermittently. My belief is that it has been continuously known from an immensely ancient source and has simply run down at different rates in different places. Although it is subject to sporadic renaissance, alas, the trend is ever downwards."*)

Athelstan, a grandson of King Alfred, was king of Mercia in 934 and the following year was elected 'king of all the English'. He codified the laws of the country and refined its standards of measure. One of these, known as the King's Girth, represented the distance from the monarch or his residence within which all offences were regarded as treason against the Crown. Its stated length was: 3 miles, 3 furlongs, 9 acres, 9 feet, 9 palms and 9 barleycorns. Of these units, the mile of 5,280 feet, the furlong of 660 feet and the foot itself are still in use today. The barleycorn was one third of an inch and the palm was 11 barleycorns or 3 2/3rds inches. The acre, which is now exclusively a unit of area, once had a linear application. According to Richard Bernese's work on land surveying published in 1537, its length was the shorter side of a rectangular strip measuring 4 by 40 perches. The present accepted value of the perch is 16½ feet, making a linear acre 4 x 16½ or 66ft [later known as a chain]. These were called field measures and were used for surveying arable land. Longer than them by 1 part in 11 were the units applied to forested land: the woodland perch of 18ft and its linear acre of 72ft. The linear acre in the King's Girth is this 72ft measure.

The length of the King's Girth is therefore:

3 miles of 5,280ft	=	15,840ft
3 furlongs of 660ft	=	1,980
9 linear acres of 72ft	=	648
9ft	=	9
9 palms of 0.305556ft	=	2.75
9 barleycorns of 0.027778ft	=	0.25

18,480 ft = 3½ miles *exactly*.

Taking this measure as a radius, the area of the King's Girth is 7 miles wide. As a square it contains 49 square miles, or 38½ if it is a circle.

This curiously detailed account of the various measures making up the King's Girth is also curiously imprecise. The Saxon lawmen make no mention of the shape of the area, whether it is a square or a circle, and we are not told whether the 3½ miles is a radius or a diameter. These omissions are significant and imply that the King's Girth is not just a measured length or area but a symbol of an ideal state. It is a symbol that relates to three different levels of reality. First, it is a cosmological pattern, expressing through number, measure and geometry, an aspect of the heavenly order. Coming down to earth, it delineates the ritual order of the Saxon court and state; and, on the material plane, it is the model of the ancient English manor system.

The basic pattern is neither a square nor a circle but both together – a circle contained by a square of 7 miles. The perimeter of the square is 28 miles, of the circle 22 miles. The square is divided into 49 smaller squares, each containing 1 square mile or 640 field acres of 43,560 square feet. This arrangement was exactly repeated in the USA, much of which was divided in the same way into sections of 640 acres.

As the ideal constitutional pattern of the royal court, the King's Girth diagram [see below] is divided into two main parts. The central square contains the royal park and residence, while the rest of the area stretches away for 3 miles on each side. This outer area is the royal domain or domestic hunting ground, well forested and stocked with game.

The central square, 1 mile wide, is divided into 8 x 8 = 64 squares, each 1 furlong wide. Sixty squares are occupied by parkland, while the central four squares contain the court precinct, the King's 24ft square court or chamber, the central platform of 6ft and the pole – 6 inches thick – that forms the axis of the whole ritual order.

Reflecting the pattern of the royal court, but on the less formal level of actual human life, is the manorial plan. The manor is a self-governing village whose main components correspond to each of the six units of measure in the King's Girth. The three smallest – the barleycorns, palms and feet – make up the central 24 ft-wide area representing the sanctuary that later became the site for a village church. The area measured by acres is the residential part of the manor; occupied by houses, workshops and gardens. Beyond that, measured by furlongs, are the cultivated fields, divided into long, parallel strips by the rod, chain and acre, and apportioned out each year between the manor families. Surrounding the fields is the greater part of the manor lands, stretching for 3 miles in each direction and consisting of meadows, commons and woodlands. These provide fuel, timber for building, pasturage for domestic animals, game and all other needs of the manor.

Here is a summary of the pattern in units of feet, with its symbolism on three levels: the heavenly, the kingly and the practical.

UNIT	Length x 2	Total width	Functions and symbolism
9 barleycorns	0.5	0.5	world-axis, eternal law, central standard
9 palms	5.5	6	seat of divine justice, throne, central rock
9 feet	18	24	hall of the gods, king's court, central sanctum
9 acres	1,296	1,320	home of the gods, court precinct, manor dwellings
3 furlongs	3,960	5,280	abode of the just, royal park, cultivated land
3 miles	31,680	36,960	Elysian fields, royal forest, commons / woodland

This is a pattern of an ideal realm with a state constitution designed to reflect the order of Creation. It is not a contrived pattern, peculiar to the Saxons or any other race or age, but derives from an ancient tradition of spiritual and scientific knowledge that has left its mark in the cultures – and most plainly in the identical units of measure – of people in every continent. Hierarchical, cosmologically based societies in NW Europe, from the Bronze Age Celts to the mediaeval Vikings, have their prototype in this same numerical pattern which, according to classical historians, is revealed to humanity by the gods at different times and places.

Everything in these societies was apparently centred upon the king who, like Athelstan, was elected as the best man among his peers. But he was merely a symbol and executor of law, not the source of it. At the heart of all is a rock or altar with the attributes of an *omphalus* or world-centre, revered by the whole tribe or nation as its birthplace and the symbol of its identity. In Athelstan's scheme its measure is 6 ft across, which is also the length of the sceptre or rod by which the king metes out or measures out justice. And through the centre of the altar runs a stout, vertical pole, six inches in diameter, that is regarded as an extension of the cosmic axis between the two poles of the universe. This central axis is the only fixed, unmoving component in the world of endless change that revolves around it, and it is therefore the first symbol of divine law that is also constant and unchanging. Its position, erect at the hub of the Saxon royal order, signifies a ritualized society, dedicated to upholding a traditional code of natural law that was revealed by the gods to their ancestors at the beginning of history.

Starting at the centre with the pole, altar and sanctuary, the pattern of the royal court – the mean term in the progression from the human to the divine level – expands in scale to reach its limit at the outer edge of the royal forest-lands, wherein criminals were judged guilty of sacrilege. But that is not the limit of its influence. For, as in all societies where the values and customs of the state capital set the tone for the provinces, the example of the Saxon court was imitated in villages and households throughout the kingdom. This was, in a sense, rule by fashion, but it was a fashion that did not fluctuate, for Athelstan's lawmen were concerned with stability rather than change. The purpose and effect of their state pattern was to maintain a psychological condition, properly called an ***enchantment***, in which ordinary life was exalted – experienced as blessed, divinely ordered and, as far as the pains of existence allow, happy. An enchantment implies chant, and the principal cause of law and order was music and festivals. A round of feast days marked the various stages in the farmer's and hunter's year; and at each of these, at the same spot on the usual day, local people sang the traditional songs and enacted the same rituals as they believed they had always done. Meanwhile, the king with his judges and courtiers progressed around the country, imitating the sun in its annual passage through the zodiac, visiting each of its twelve sections for one month in the year, delivering justice and upholding standards of culture and husbandry. At every spot where the royal court located itself, the sanctuary of the King's Girth was extended into the surrounding countryside.

The diagram that illustrates the King's Girth [see below] is not just a clever contrivance of numbers and measures, but gives insight into the science behind ancient statecraft. It was indeed a science, properly so called, based on the demonstrable truths of number and geometry rather than on expediency and opinion. For those who can see Beauty in her abstract, essential form, through the harmonies of number and traditional measure, many delights and wonders can be found among the areas and ratios in the King's Girth diagram. And those who have followed the discoveries of classical Pythagorean science in the designs of megalithic circles can perceive a tradition that has persisted in the British Isles from Stone Age antiquity through Saxon times and into the Middle Ages. This tradition, and even the fact of its existence, is scarcely recognized today. But those who wish can study it, to the benefit not only of themselves but of their entire community and (human) race.

The conclusions that ensue from analysis of the King's Girth are several and radical – not altogether in accord with modern notions and prejudices. The Saxons were not just rough and boorish. Their rulers were learned men in the classical sense; educated in number, measure, geometry, astronomy, land surveying and the classical sciences. Behind that level of education lies another, relating to a form of science more subtle and valuable than any that is known today. The evidence of archaeology and anthropology is unanimous; that the ultimate purpose of this further science was for divination – for communication with local spirits and those of the dead, and for attracting grace, divine blessings or plain good luck to one's self and surroundings.

The King's Girth exemplifies this science. It depicts a little universe, self-contained and with its own laws, customs, music and traditions. Whether it is a kingdom, province, manor or an individual household, its basic pattern is the same, and this pattern is not only expressed in land measures. It lies behind every institution and influences the whole of society, including its religion and world-view. In archaic England, from pagan into Christian times, the centre-point of every judicial or legislative assembly was a tree, rock or upright pole. Around it was an open space, fenced off and dedicated to the supreme deity who upheld the principle of free speech within the sanctum. It was a place of ceremony, legislation and justice. Democracy has its natural roots there, for everyone on the manor attended its court and was party to its legal and administrative proceedings.

An essential feature of this social and constitutional order is its numerical basis. In classical, civilized societies the dominant number in state and religious institutions is Twelve, representing the 12 gods in the zodiac, 12 solar months and many other examples of twelve-fold order. The values symbolized by this number are solar, rational and imperial. Seven, on the other hand, is the natural number of the soul and spirit, of oracles and inspiration. It is the characteristic symbol of archaic and nomadic societies, where the 7 planetary gods, the 7 stars of the Great Bear and the 7-day week of the moon dominate life and ritual. The relationship between these two number symbols is illustrated in music; seven being the number of notes in the simple music of the shepherd's pipe, whereas in the sophisticated scale of religious chant the number is twelve.

Seven, the number of miles in the overall width of the King's Girth, is indicative of a very ancient measure, not of classical origin but a survival from the old Nordic culture that gave birth to the Celtic, Viking, Saxon and other traditions. Yet behind this apparent Nordic source is something greater, a unified code of knowledge that can be recognized in the primordial traditions of all nations. This leads to the big question in ancient history. Was there once, many thousands of years ago, a universal culture, embracing all humanity? Many speculative writers today are thinking in that direction, but the evidence is against it – one objection being that the origins and heydays of every nation and culture are in different periods of history. There seems to have been a succession of cultural revelation – a more interesting conclusion because it implies that revelation is an ever-active principle, as much so today as at any other time. Basically it is a code of law – not a legalistic or moral code nor a set of precepts but a numerical pattern – an abstract expression of the universal culture. It is that constant, divine law which is symbolized by the unmoving axis of the cosmos and the upright pole at the centre of the Saxon manor.

So the quaint old measure of the King's Girth actually introduces us to the educated Saxon mind, to the religiously ordered world-view established in it and to the corresponding order of society. Further pursued towards its roots, it leads us through knowledge of ancient science to confront the most serious question of all. Is this world an artefact? The premise of the question is this. The noble Saxons, in common with the priests, initiates and philosophers of all antiquity, understood the world to be a divine creation, perfectly composed in every detail of number, measure and weight. That is the traditional, ever-recurring, Platonic view of things. It occurs spontaneously to anyone who studies the ancient sciences and the subtle, comprehensive code of number that underlies them all. And it is now occurring in modern physics. According to the established Anthropic Principle, it so happens that the balance of forces in the universe is, against all conceivable odds, exactly as it would have to be to allow for intelligent life. To many physicists this looks like intelligent design. There is only one self-conscious form of intelligence that we know of – ourselves. But since we cannot claim the credit for having created the universe, we have to acknowledge another, higher intelligence. That is how we recognize a Creator. *And that is why they had a six-inch wide pole at the centre of the Saxon manor.*

1 The area of the King's Girth, whether regarded as a square or a circle, measures 7 miles across. This diagram shows the overall plan. The square is divided into 7 x 7 = 49 smaller squares, **each of 1 mile.** The distance from the side of the shaded squares to the side of the outer square is 3 miles, the 1st measure of the King's Girth.

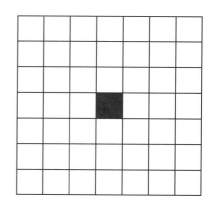

2 This represents the central square, 1 mile wide, in the previous diagram. It is divided into 8 x 8 = 64 smaller squares, **each of 1 furlong (660 ft).** The distance from the side of the central, shaded block of squares to the side of the greater square is 3 furlongs, the 2nd measure of the KG. The 3rd measure, 9 linear acres of 648 ft, lies within and occupies most of the central shaded area. At the very centre is the 24 ft square of the central shrine, the 6ft square of the central altar and the 6in wide pole at tahe hub of all.

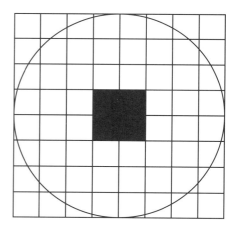

3 So here is the tiny 24 ft-wide square at the centre of the previous diagram, divided into 8 x 8 = 64 smaller squares, **each of 3 ft or 1 yard.** It contains, shaded, the 6 ft altar and the central 6in wide pole.

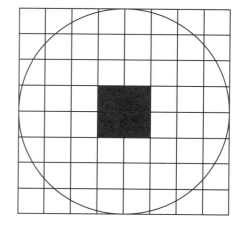

Select Bibliography

Tables of Physical and Chemical Constants (16th edition) Kaye and Laby: Longman 1995 [this remains the 'Bible']

For Good Measure William D Johnstone: NTC Publishing Group, Chicago, 1998 ISBN 0 8442 0851 5

The Sizesaurus Stephen Strauss: Kodansha America Inc, New York, 1995 ISBN 1 56836 110 6

How Heavy, How Much and How Long? Colin R Chapman: Lochin Publishing, Dursley, GL11 5RS, 1995 ISBN 1 873686 09 9

A Beginner's Guide to Stone Circles Robin Heath: Hodder & Stoughton, 1999 ISBN 0 340 73772 7

The Dimensions of Paradise John Michell: Thames & Hudson, 1988 ISBN 0 500 01386 1

The Ascent of Man Jacob Bronowski: BBC, 1973 ISBN 0 563 104988

Opus 2 – All Done with Mirrors John Neal 2000 ISBN 0-9539000-0-2
johnneal@secretacademy.com

Megalithic Mathematics (working title) Anne Macaulay: (pending) 2000

Weights and Measures, origins and development in Great Britain up to 1855
F G Skinner: HMSO, 1967

The Weights and Measures of England R D Connor: HMSO, 1978

Dictionary of English Weights and Measures R E Zupko: University of Wisconsin Press, 1968

Historical Metrology A E Berriman: Dent, 1953

Emanuel Raft Peter Pinson: Craftsman House (Sydney), 1997 ISBN 90 5703 52 19